MANHATTAN BRIDGE

the troubled story of a New York monument

Thomas R. Winpenny

Published by Canal History and Technology Press
National Canal Museum
Hugh Moore Historical Park and Museums, Inc.
30 Centre Square
Easton PA 18042-7743

in association with the Smithsonian Institution

ISBN 0-930973-29-1
Library of Congress Control Number: 2003111759

Canal History and Technology Press
National Canal Museum
30 Centre Square, Easton, PA 18042-7743

for my beloved Marilyn

contents

acknowledgments

This book grew very logically out of my earlier *Illustrated History of the Phoenix Bridge Company* of Phoenixville, Pennsylvania. Phoenix Bridge designed and fabricated wrought-iron truss bridges and suspension bridges for erection all over the world from Needles, California, to Vladivostok, Russia. While pouring over the records of Phoenix Bridge in the Hagley Archives, I noted extensive entries for fabrication for the Manhattan Bridge. This project was very important to Phoenix Bridge because it was a much-needed major achievement in 1909 in the wake of a horrible disaster involving the company in Quebec in 1907. In short, the success of Manhattan Bridge atoned for the failure of Quebec Bridge and thereby redeemed the Phoenix Bridge Company.

Why not a book on Manhattan Bridge? Surely the American public had overdosed on analysis of the beloved Brooklyn Bridge by now. Why not take a look at the adjoining span that was designed to relieve the congestion and near-rioting on the Brooklyn Bridge at rush hour? Perhaps in a perfect world my friends and colleagues would have rallied round and provided encouragement. Essentially, what they did was smile politely. Nevertheless, the project moved ahead during the late 1990s. Elizabethtown College contributed a sabbatical to aid in research and writing; and again the Hagley Archives, presided over by Chris Baer and Marge McNinch, provided much of the critical material. The newspaper of record, *The New York Times*, was carefully gleaned by my research assistant, Amy Kohut, also funded by Elizabethtown College.

The engineering fraternity was helpful in answering questions—from Herbert Rothman at Weidlinger Associates in New York (responsible for restoration of the bridge) to John Kulicki at Modjeski and Masters in Mechanicsburg, Pennsylvania (responsible for the safety of the subways on the span); from Eric DeLony, Director of Historic American Engineering Record, National Park Service, in Washington, D.C., to Professor David P. Billington of Princeton University; from Jagtar Khinda, currently analyzing the Manhattan Bridge on site, to Emory Kemp, Director of The Institute for the History of Technology and Industrial Archaeology in Morgantown, West Virginia.

I am grateful for many conversations with friends and associates who know more about America's largest city than does the author. Accordingly, Paul Dennis of Elizabethtown College, Al Gillis of IBM, David Schuyler of Franklin and Marshall College, James Rasenberger of Manhattan, and Cyril J. Smith of Garden City, Long Island, provided much-needed light where only darkness had existed.

This bridge book also profited from the work of talented editors and illustrators Ann Bartholomew and Lance Metz of Canal History and Technology Press. Jack Hedrick of the Elizabethtown Chemistry Department served as a valuable and valued first reader.

I am indebted to Professor Eugene S. Ferguson, formerly of Iowa State and the University of Delaware, who showed me that scholars trained in the humanities really did have something to say to engineers.

Finally, my greatest "thank you" in any project goes to my wife, Marilyn Winpenny, who finds time to run our house, teach gifted children, counsel teenagers in drug rehabilitation, and provide love and spiritual advice to the author.

Tom Winpenny

foreword

Manhattan Bridge, a Model for Deflection Theory

Eric DeLony

Chief, Historic American Engineering Record

National Park Service

Many historians consider the Brooklyn Bridge (1883) one of the greatest technological achievements of the nineteenth century. Consequently, the Manhattan Bridge (1909), located adjacent to its more famous neighbor, has always been in its shadow.

Despite the bridge's perceived lack of distinction, Tom Winpenny's book reveals that the Manhattan Bridge represents a significant step in the evolution of world bridge-building technology. Manhattan Bridge was a seminal step in the theory of structural engineering. Designed by Leon Moisseiff (1872–1943), it was the first major bridge built solely within the twentieth century.[1] Despite its Beaux Arts embellishments, Manhattan is the origin of contemporary suspension-bridge design, influencing subsequent suspension bridges to this day. The "new" deflection theory on which its design was based quickly supplanted the older approximate-analysis method. Deflection theory revolutionized the way bridge engineers designed suspension bridges for the rest of the century. Manhattan Bridge represents this important theoretical advance.[2]

While the new theory took into account the inherent stiffness of the cables from the axial forces produced by the dead load, providing a method to accurately assess the interaction between the main cables and the stiffening trusses, it did not take into account the problem of the subway trains on the outer limits of the lower deck. For reasons unknown, Moisseiff placed the heaviest live loads, the subway trains and elevated-car traffic, outside rather than inside the towers, resulting in excessive torque on the outermost edges of the bridge. This has resulted in perennial problems. Over the years, heavy subway traffic has caused extensive twisting of the deck. Maximum torsion occurs when subway trains start to cross from opposite ends of the bridge at the same time. At that moment, one side of the roadway dips four feet to the north, while the other dips four feet to the south, resulting in a total roadway deflection of eight feet.[3]

Due to advanced theoretical understanding developed since Manhattan Bridge, it is recognized that the bridge was underbuilt. According to the analysis of Weidlinger Associates, the engineers of record for the current rehabilitation,

> Manhattan Bridge has always been subject to large movements resulting in massive cracking. Manhattan was first designed by deflection theory ... available technology did not allow for the theory to be applied with the sophistication and detail required by the bridge's complex structural system.[4]

The present rehabilitation is designed to correct the structural deficiencies based on the increased knowledge and understanding of deflection theory. It has been underway for over a decade, and it is estimated that it will take another several years to complete the work by 2008. The total estimated cost is $702 million. The bridge when originally constructed cost $17 million.[5]

Consulting engineer Leon Moisseiff was the innovator who introduced this new way of looking at suspension bridges to America. Manhattan's suspension span (1,470ft) is less than either of its two predecessors, Brooklyn and Williamsburg, but it marks a fundamental leap over the previous structures, and contains many of the seeds which blossomed in subsequent suspension bridges for the rest of the century. Built between 1901 and 1909, this 6,855ft-long suspension bridge system connects Manhattan with the Borough of Brooklyn. The bridge touches down on Chinatown's Canal Street in Manhattan and Flatbush Avenue in Brooklyn. An estimated 78,000 vehicles cross the bridge daily. Like the other three East River bridges, it allows access by pedestrians, bicyclists, automobiles and is one of three Manhattan bridges to carry subway trains — the B, D & Q lines.

While Moisseiff was the one who crunched the numbers and finalized the calculations, Othniel Foster Nichols was the chief engineer, appointed by the City of New York to oversee construction. Nichols was the one who approved a design based on a new theory — deflection theory — that was first developed by Austrian engineer Josef Melan and used for the first time on a major suspension bridge by Moisseiff, a design engineer on the staff of the bridge department. Deflection theory recognizes that the dead load significantly moderates the distortion under live load. The

View over the East River toward Manhattan, with the Williamsburg bridge in the foreground, the World Trade Center in the background, and the Manhattan and Brooklyn bridges spanning the river in the distance. (Photograph by Jack Boucher for Historic American Engineering Record, 1978)

immediate consequence of this exciting new way of evaluating the behavior of suspension bridges was the realization that stiffening trusses, common to all bridges up to this time, had little to do with reducing deflections on large suspension spans. Deflection theory suggested that the depth of the truss could be reduced without impairing the stiffness of the bridge. This allowed a drastic reduction in the size of the deck stiffening members, resulting in a sleek, elegant, ribbon-like look.[6] It rationalized the direction of suspension bridge design from the Manhattan Bridge through the era of American dominance of suspension bridge design into the 1960s.

The theory was introduced in 1888 by Austrian engineer Josef Melan, who first published it in *Handbuch der Ingenieur-Wissenschaften,* 2nd edition, Leipzig. The first translation of Melan's theory into English appears in an appendix to the 1894 US Army Corps of Engineer's report, *Maximum Span Practicable for Suspension Bridges.* Deflection theory also appears in the 1910 edition of Johnson, Bryan and Turneaure's *The Theory and Practice of Modern Framed Structures*, a book widely used as a text in engineering schools at the time. In 1913, David Steinman published a complete translation of Melan's theory from which he developed his own *Practical Treatise on Suspension Bridges* in 1922. In 1929, Steinman included deflection theory in the new edition of his book.

Melan, as described in Steinman's book, laid the foundation for deflection theory by accounting for the change in the shape of the main cable under live loads. In 1909, Leon Moisseiff used Melan's formulas to design the Manhattan Bridge. It was Melan's work that led to the application of the deflection theory design in America; his early studies were the means by which these theories were adopted by American engineers in the early twentieth century.[7]

By the second decade of the twentieth century engineers were well aware of the economy that could be achieved using deflection theory. In the twenty years following completion of the Manhattan Bridge, three record-breaking suspension bridges were built: Bear Mountain on the Hudson River in 1924, 1,632ft; Delaware River in Philadelphia in 1926, 1,750ft; and the Ambassador in Detroit, in 1924, 1,850ft. The economy of deflection theory allowed for the span-length to truss-depth ratio for these bridges to decrease, resulting in ever-shallower trusses that used less material. The aim of suspension-bridge design in the late 1920s and the 1930s was to decrease the importance of the truss as a stiffener, resulting in smooth, clean, sharp lines, a shallower sinuous ribbon of plate girders, devoid of extraneous architectural embellishments.[8] While the primary objective was efficiency and cost savings, this new mathematical model also allowed engineers to design suspension bridges more elegantly.

Looking south toward the George Washington Bridge over the Hudson River. (Photograph by Jet Lowe for Historic American Engineering Record, 1986)

The next giant step forward in suspension bridge design occurred twenty-three years after Manhattan when the George Washington became the longest suspension bridge in the world, doubling the clear span from 1,470ft to 3,500ft. To achieve the record length, chief bridge engineer Othmar H. Ammann had to rethink conventional parameters of bridge design. While engineers had resolved the effects of live loads of heavy locomotives during the nineteenth century, the advent of motorized vehicles necessitated a whole new set of calculations because of the uneven movement of the deck and cables due to the weight of heavy trucks and lighter cars moving across the bridge. A unique feature of the design by Othmar Ammann was the lack of stiffening trusses. Ammann conceived this design in consultation with design engineer Leon Moisseiff. Both figured that the great width and weight of the deck gave it more than enough stiffness.

It should come as no surprise that the justification for reducing the stiffening truss was found in the "deflection theory" that had been used by Moisseiff to analyze the Manhattan and Ben Franklin spans.[9] Moisseiff wrote in 1939 that the George Washington Bridge, with its unstiffened deck, brought home the seeming paradox that the cable is the stiffest element in a suspension structure. Ammann confessed that there was no doubt in his mind that a span of 3,500ft, with an ultimate dead weight of 15 to 20 times the greatest wind force exerted vertically on the deck, could not be affected by the wind.[10]

Moisseiff, who was the consulting engineer on this bridge along with chief engineer Othmar Ammann, discovered that the tremendous dead load, 39,000lbs/linear foot, caused the cable to be the stiffest member in the bridge. When the second deck was added, the depth of the stiffening truss of the George Washington bridge was only 29ft or 1/120 of the span. The stiffening truss of the Golden Gate was even shallower at 25ft or 1/168 of the span. Making the stiffening trusses of less depth allowed engineers to abandon the open fretwork of deep stiffening trusses reminiscent of nineteenth-century bridges. Plate girders presented a clean and simple line, were easier to fabricate, easier to connect, and easier to maintain. The Bronx-Whitestone (1939) and Tacoma Narrows (1940) bridges were the last two legacies of the Manhattan Bridge. These two bridges pushed deflection theory to its limit, and beyond the understanding of engineers. The Whitestone had plate girders only 12ft deep while the depth of the plate girders on Tacoma Narrows was reduced to 8ft. The George Washington began a decade of bridge design during which seven bridges with spans in excess of 1,000ft (305m) were constructed in the United States, ending with the dramatic collapse of the Tacoma Narrows Bridge in 1940.[11]

As big as the George Washington was, the giant in all categories of bridge design was the Oakland Bay Bridge (1936). Its truss, tunnel, and suspension system extended for eight miles in two parts separated by Buena Vista Island. The East Bay crossing is a succession of cantilever

Oakland Bay suspension bridge, seen from Yerba Buena Island. The central anchorage is in the middle. (Jet Lowe photograph for Historic American Engineering Record, 1985)

trusses while the West Bay is an unusual double suspension span whose design was the next to be based on deflection theory.

Founding the central common anchorage for the paired suspension spans required the world's largest caisson, 92ft x 192ft. This was sunk to a depth greater than 200ft below low tide, in over 100ft of water with swift tidal currents. Oakland Bay Bridge was built under the direction of C.H. Purcell, chief engineer of the California Highway Department, Glenn B. Woodruff, design engineer, with a team of consultants: Ralph Modjeski, Charles Derleth Jr., and Leon Moisseiff—the best talent available.

The next bridge whose design was based on deflection theory was the Golden Gate (1937). All bridge engineers dream of the opportunity of building the ultimate span and the Golden Gate on the northern coast of California provided the perfect opportunity. Part of the dream is finding the ideal location where the finished bridge stands out in its monumental glory for all to see. Few bridges capture this ideal better than the Golden Gate. Located in sublime isolation on the northern tip of the San Francisco peninsula, the city required more efficient access than ferry boats to link it with its burgeoning suburbs on the other side of the bay. Spanning the Golden Gate seemed a dream rather than reality. A water body this wide having never been bridged required "the world's greatest bridging enterprise," conceded Joseph B. Strauss, the bridge's engineer. While Strauss clearly was ahead of his time in engineering, marketing, and dealing with chambers of commerce, he spent nearly two decades politicking, promoting, and constructing the Golden Gate Bridge. Charles Ellis is credited with working out the mathematical design.

Strauss's original design—part suspension, part cantilever—was rejected by a panel of engineers.

In 1922, Strauss hired Charles Alton Ellis, an unassuming professor of structural engineering at the University of Illinois who had formerly worked for the Dominion Bridge Company of Chicago, as his site engineer. He was charged to develop the new design. Ellis arrived at a Moisseiff-inspired suspension bridge implementing Moisseiff's deflection theory. The mathematics for the design filled eleven and a half volumes.[12] When completed, Golden Gate was the longest suspension span in the world at 4,200ft and remains, today, a symbol of the city.

Golden Gate Bridge, looking north, showing the "bay side" of the structure. (Jet Lowe photograph for Historic American Engineering Record, 1984)

The last bridge whose design was derived from the Manhattan Bridge is the first Tacoma Narrows span. In this case Moisseiff, who was recognized as the leading American theoretician of computational analysis of suspension bridges, was the design engineer rather than the consultant. With a 2,800ft suspended span, it was the third-longest suspension bridge in the world and represented the ultimate fulfillment of design goals to make the decks of suspension bridges ever thinner. Stiffening girders were only eight feet deep, or 1/350 of the span. On November 7, 1940, four months after its opening, the thin girders of Moisseiff's Tacoma bridge lay in the river, brought down by a steady 40mph wind. Though no lives were lost, this was the twentieth century's most spectacular bridge failure. A baffled Moisseiff was quoted to be "completely at a loss to explain the collapse."

The Tacoma collapse took with it a generation of suspension-bridge engineering tradition,

Historic view of Tacoma Narrows Bridge, 1940, looking west across the toll plaza. (Historic American Engineering Record)

but out of this failure would emerge the new science of bridge aerodynamics, yet another new form of suspension-bridge design. The dynamic forces of wind proved to be the undoing of a generation of engineering thinking.[13] Twentieth-century engineers adopted Melan's deflection theory because it provided great economy, especially in the deck stiffening. So great were the savings, it was possible to eliminate deck stiffening altogether. This trend toward thin, ribbon-like decks led to wind-induced undulations and eventually to the Tacoma failure.[14]

This bridge collapsed, only four months after opening, under the action of 42 mph wind-induced oscillations—one-third the design pressure of 120 mph. This most spectacular bridge failure of the twentieth century exposed over a century of ignorance of the dynamic effect of wind on bridges. The wind that destroyed the Tacoma Narrows Bridge generated an oscillating motion in the supple deck, which gave way over time. Aeroelastic instability, or flutter, caused the moving structure to generate its own aerodynamic forces, especially twisting, and ultimately to fail due to the aerodynamic instability caused by the lack of sufficient deck stiffening. Following the collapse, the Bronx-Whitestone Bridge, another bridge based on deflection theory, was quickly reinforced, first with a system of stays and then with a truss to further stiffen the slender girder deck.

The dramatic failure of Tacoma Narrows, filmed in motion pictures, is shown to civil engineering students throughout the world and resulted in the complete reevaluation of suspension bridge design. During the ensuing investigation, engineers drew from theoretical concepts such as lift, eddy formation, vortex shedding, negative damping, and flutter, which were developed by aeronautical engineers for designing airplane wings but are applicable to the design of suspension bridges.[15]

F.B. Farquharson, in his report on the collapse, wrote that "the collapse ... came as such a shock to the engineering profession that it is surprising to most to learn that failure under the action of wind was not without precedent in the history of suspension bridges."[16] The collapse of Samuel Brown's Brighton Chain Pier in 1836 provided the first clear evidence of resonance. Thomas Telford's 1826 bridge across the Menai Straits provided the most compelling evidence of aerodynamic instability. In 1839, the bridge was damaged during a storm. Observations of the effect of wind on Telford's bridge were that even though the wind blew nearly perpendicular to the side of the structure, it produced no lateral motion, only vertical undulations. This was early recognition that transverse, horizontal loads excited vertical undulations along the length of the bridge rather than sideward swaying. John Roebling was the only engineer of his era who appreciated the true nature of the phenomenon — intuitively, not scientifically — and attempted to stabilize his bridges with stiffening trusses and stay bracing. Elastic theory, introduced in the late nineteenth century, emphasized deep trusses to support the heavier and fast-moving live loads of trolleys and subways. Roebling's stays disappeared because of calculation difficulties with indeterminate stresses. Beefed-up trusses solved the problem of aerodynamic instability, but the renaissance in modern bridge engineering wrought by Moisseiff and deflection theory led to longer, lighter and more flexible vehicular-only bridges. Twentieth-century engineers had no direct memory of aerodynamic failure.[17]

In late March 1941, when Ammann, von Karman, and Woodruff submitted the final report on the failure of the Tacoma Narrows Bridge, they concluded that the crucial event in the collapse was the sudden change from a vertical to a torsional mode of operation. The design erred in creating a structure with an extraordinarily small width compared to its span. Their expectation that the bridge would have adequate lateral rigidity under wind pressure was unquestionably justified, but the combination of the narrow width with great flexibility proved to have been responsible for the excessive torsion under aerodynamic action that led to failure of the floor. This experience reveals the danger of excessive narrowness. The fundamental weakness of the Tacoma Narrows Bridge was its great flexibility, vertically and in torsion. The comparatively narrow width gave adequate lateral rigidity, but combined with the extreme vertical flexibility, the narrowness made the structure extremely susceptible to torsional motions created by aerodynamic loading. The resistance of a suspension bridge to vibrations from dynamic forces is furnished by its stiffness, inertia and structural damping. The Tacoma Narrows Bridge was deficient in all three respects.[18]

In the six decades following Tacoma, suspension-bridge design has been dedicated to the interaction between moving air and structure and to understanding its various guises. Although von Karmen's "vortex street" was known by aerodynamic engineers, the existence of vortices had never been considered by structural engineers. Unfortunately, these forces were in exact resonance with the twisting oscillations of suspension bridges. Aerodynamic stability is a fickle phenomenon. The understanding of it remains incomplete.[19]

The final word on the Tacoma Narrows' collapse and a word of caution to future engineers was given by the eminent historian of engineering James Kip Finch, who also was professor of civil engineering at Columbia University. He provided an astute introspective observation of Tacoma's infamous failure and on deflection theory by recounting the wind-induced collapses of Telford's Menai Straits bridge in 1839 and Charles Ellet Jr.'s Wheeling bridge in 1854. He stated that these difficulties with earlier bridges were long forgotten. To modern engineers the gyrations of Tacoma Narrows constituted something entirely new and strange. They were not new. They had simply been forgotten.

Museum of the City of New York

chapter one

MANHATTAN BRIDGE

Introduction

The modern American landscape is punctuated by a host of magnificent bridges, from the Brooklyn Bridge in the East to the Golden Gate in the West. Justifiably, many of these structures are admired as works of art and engineering genius. They are revered as national treasures and sources of great local pride. The artistic are moved to paint them, photograph them, compose songs and write poetry about bridges. Lovers stroll hand-in-hand across these spans, desperate people jump, and the homeless find imperfect refuge beneath.

Most great bridges constitute one more American success story. Everyone is delighted to drink in the visual pleasure they provide, motorists and subway strap-hangers alike appreciate the vista while crossing, and taxpayers understand that they are cheaper to build than tunnels.

Beyond these wonderful generalizations, however, a darker side lurks. Much of this darker side stems, historically, from a lack of basic maintenance, and a few extremists who contend that major bridges cause poverty and depressed land values. Bridges, like almost everything else, require a bit of care and looking after. Local governments, sometimes with federal aid, ideally, invest in painting, fighting corrosion, and inspecting for cracks and other structural problems. In practice, of course, maintenance is frequently deferred and deferred and deferred—until safety considerations finally override budgetary limitations. The net result today is that many of America's bridges are in serious need of major repairs and restoration.

Having noted the general societal inclination to not care for its bridges, what happens when we couple this societal predisposition with original design flaws and bad engineering choices? Enter Manhattan Bridge. The saga of this span is fascinating for at least three reasons. First, similar to the Brooklyn Bridge that took fifteen years to complete (1869–1883), the Manhattan also got caught in a tangled web of urban politics and required roughly ten years to complete. Second, this East River structure, built between 1900 and 1909, was conceived at a time when bridge engineering theory was in flux. As Thomas Kuhn in *The Structure of Scientific Revolutions* would state the matter, they were shifting from one "scientific paradigm" to another (discussed in "Appendix").[1] Unfortunately for the Manhattan Bridge, the newer paradigm was not fully refined until the 1920s. Third, a dismal maintenance schedule superimposed on a heavily used bridge with design problems was a perfect formula for disaster. Not a literal disaster in which portions of the span crashed into the East River (actually an estuary), but a figurative disaster in which repairing, reconstructing, and maintaining the Manhattan Bridge became a black hole capable of devouring hundreds of millions of dollars, with only modest results (some New Yorkers are quick to point out that the city has never suffered a major bridge disaster).

Indeed, the Manhattan Bridge and its constituency—primarily commuters traveling between Brooklyn and Manhattan on weekdays and "day-trippers" to Coney Island on the weekend—became analogous to two dependent people stuck in a bad marriage. In one sense they could not afford to stay together, and yet they could not afford to be apart. Pouring hundreds of millions of dollars into rebuilding the Manhattan Bridge seems to make little economic sense, and yet the quarter-million riders on the subways alone that use the structure cannot imagine life without it.

This book endeavors to capture the unvarnished (or should it be "unpainted"?) truth regarding the history of Manhattan Bridge, arguably a thankless and unprofitable task as much of the reading public prefers success stories. After all, as the distinguished historian David McCullough demonstrated not too long ago, there is always room for one more scholar obsessing over the Brooklyn Bridge, whether presented as a book or video tape.[2] By contrast, the subject of this study is a neglected stepchild, rarely deemed an artistic triumph and surely not serving "the better part of town" in either Manhattan or Brooklyn.

Design credit is normally divided among Gustav Lindenthal, Leon Moisseiff, and O.F. Nichols. The architectural firm Carrere and Hastings of New York City designed the portals. Essentially, the superstructure known as Manhattan Bridge is a modern agglomeration of steel and nickel steel fabricated in Phoenixville, Pennsylvania, by the Phoenix Bridge Company, accented with zinc-coated suspension cables, the sort generally provided by the Roeblings of Trenton. A cynic might protest that it has all the charm of an overly modern structure (in 1909) created by a hyperactive twelve-year-old with a gargantuan erector set. (Alas, erector sets had not yet been invented when this span was put up.) Perhaps writing this book has something to do with loving the unlovable. Maybe the author suffers from a fixation on bridges and thus almost any subject will do. Enough philosophizing.

The opening decade of the twentieth century, the period of design, steel fabrication, endless delay, construction of piers, and erection of the superstructure, is chronicled in the next chapter. It focuses on the entanglements of machine and reform politics, squabbling engineers, and a wide variety of additional actors; the theme that emerges is that "business as usual" in Gotham proved to be terribly costly and resulted in delays so extreme that observers justifiably labeled them scandalous. The public interest or the teamsters, motorists, and "straphangers" lost again. Fortunately for scholars, *Engineering News*, *Scientific American*, and *The New York Times* faithfully recorded and thus preserved the melodrama.

The labor scene is addressed in the third chapter, "The High Steel Worker's World." What did it mean to do high steel work? What sort of folklore surrounded the labor scene? What do injury and fatality rates look like? Who should take responsibility for the human carnage? What difference does a union make? What can be learned from *The Bridgemen's Magazine*, the monthly voice of the International Association of Bridge and Structural Iron Workers published in Indianapolis? How did work on the Manhattan Bridge fit in with the famous or infamous dynamiting strike conducted by the IAB&SIW during the first decade of the new century? Professor Sidney Fine has produced an excellent study of the struggle between the International Association of Bridge and Structural Iron Workers and the National Erectors' Association or bridge fabricators, entitled *Without Blare of Trumpets*.[3] Why did the Phoenix Bridge Company agree to let this East River span be a union job (with a 30-foot union banner) when the firm had a long history of being overwhelmingly committed to the open shop crusade? What were the sources of tension between Phoenix Bridge and the Ryan-Parker Construction Company of New York?

The fourth chapter, which covers the first forty years of the bridge linking Manhattan's Lower East Side and South Brooklyn, takes a closer look at the engineering debate, something that went far beyond mere theoretical discourse during the period 1906–1909 when the Manhattan Bridge superstructure was actually being hoisted into place. The collapse of the Québec Bridge over the St. Lawrence River in August of 1907, killing 75 workers on the south arm, was more than just one more grotesque newspaper story. The human carnage and twisted steel reminded readers in both the United States and Canada that, indeed, great bridges still collapse. What's more, the Phoenix Bridge Company that, arguably, had created the Québec disaster, was busily at work on this major project over the East River. Debate raged on such topics as: Did American civil engineers really know what they were doing? What is "deflection theory," and why was it so attractive? What is a paradigm shift? Why not recreate the glorious masonry towers of the Brooklyn Bridge? Who in their right mind ever decided that subway trains (each empty car weighing forty tons) should run on the <u>outside</u> lanes of Manhattan Bridge—or any other bridge for that matter?

The fifth chapter is in some ways the key to this entire book as it offers an overview of the serious structural problems that surfaced in the 1950s. These problems were so severe that some observers considered abandoning the Manhattan Bridge—deeming it a "lost cause" after little more than forty years of service. The difficulties stemmed from flawed design and the frequent absence of maintenance.

A careful combing of the historical record documents some traces of major concerns as early as the late 1930s. The talk of abandonment in 1953 sounds absurd to any civil engineer or serious bridge historian. Knowledgeable observers cannot help but be mindful of the iron bridge erected across the Severn River in Ironbridge, England, in 1779 that currently is still standing, though no longer in use for vehicular traffic. If someone wants to argue that the span at Ironbridge is an exception, the fact remains that almost any wrought-iron truss bridge of the nineteenth century should serve effectively for at least a century—with any maintenance at all. At least, this is the position of Eric DeLony, Chief of the Historic American Engineering Record (HAER) and one of our nation's leading bridge historians. (Bridge designer Gustav Lindenthal once observed that with proper maintenance the Manhattan Bridge might last forever.) Having noted all this, through the second half of the twentieth century all conceivable problems seemed to be compounded. The flow of traffic increased almost beyond belief, the resulting torque on the outside of the span created by endless subway trains was extraordinary, and the search for corrosion only led to the discovery of corrosion under layers of corrosion! Chapter Five makes it abundantly clear that all of the problems that might have been anticipated made their presence known.

The discussion in the sixth chapter, "Tapping the Faucets in Washington, D.C.," is self-evident, given the title. The emphasis is on the reconstruction work of Weidlinger Associates, consulting engineers headquartered in New York City, work that has been underway since the late 1970s and has cost hundreds of millions of dollars. Let the record show that spending hundreds of millions in one decade on the Manhattan Bridge does not mean that a period of modest spending is in the offing. The fact is that discussions about bridge reconstruction seem to dwell on just two concepts: "decades" and "hundreds of millions of dollars." Not surprisingly, since The Federal Bridge Replacement Act was passed in 1978, the federal government has paid for eighty percent of the major reconstruction projects, leaving the denizens of Gotham to shoulder only the remaining twenty percent. With this cozy arrangement, perhaps the lights of Manhattan Bridge will continue to illumine the New York night for some time.

The seventh chapter serves as a conclusion and essentially asks, what are the insights to be garnered from this tale of woe? What does this saga tell us about the history of civil engineering in America? What does it tell us about deflection theory and paradigm shifts in design? To what extent were the problems of building and maintaining Manhattan Bridge peculiar to a single structure? Would it have been wise and prudent never to have allowed subway trains to use the bridge?

In light of the 2,000-plus repairs made annually in the early 1990s, is there any point at which the New York City Transportation Department (which owns the bridge), other city officials, the Metropolitan Transit Authority, the Brooklyn Rapid Transit Company, the commuters, or citizens turn their backs on the structure and consider abandonment? Tom Kelly, spokesman for the MTA, in the September 11, 1997, issue of *The New York Times*, offered the cryptic observation that his agency "was not considering abandoning the bridge altogether."[4] What a ringing endorsement!

Fortunately, the grand agglomerations of steel known as bridges are not readily nor casually abandoned. In this case, citizens grow accustomed to the outline of each particular span in the East River landscape. Photographers derive endless delight from grouping bridges in the same photo or capturing the outline of one through the cables of another. Indeed, living with the foibles and shortcomings of the Manhattan Bridge is simply part of being a New Yorker.

chapter two

THE ECONOMICS OF BRIDGE BUILDING

or, Why Bridges Frequently Cost Much More Than They Need Cost:

The Manhattan Bridge as a Case Study

At the turn of the century Manhattan, Brooklyn, the Bronx, Queens, and Staten Island were consolidated into America's largest city. The burgeoning population harbored rising expectations regarding the transportation network that would make the consolidation more than symbolic. The perceived need to readily visit family in another boro, socialize with friends, shop, and get to work seriously overtaxed the existing network. Rush hour, for example, created a great crush and often near-riots on the Brooklyn Bridge, which had served New York since 1883.[1]

The solution to the problem did not reside in creating more ferry lines, or waiting for the East River to freeze in winter so that citizens could walk, ice skate, or take a sleigh ride across. The answer was found in considering the construction of tunnels and more bridges. To that end the Williamsburg Bridge was completed in 1902, and the Manhattan and Queensboro bridges were to be completed sometime in the same decade. This would ultimately mean four great bridges spanning the East River, connecting Manhattan with Brooklyn and Queens.[2] Could the citizenry and merchants have asked for more from their Mayor, Bridge Commissioner, and Board of Aldermen?

In a better world than the one in which we live, a new bridge would grow out of the work of a distinguished designer such as a Roebling, a Gustav Lindenthal, an Othmar Ammann, or a Ralph Modjeski—aided by bridge engineers such as J.A.L. Waddell, Theodore Cooper, Adolphus Bonzano, or John Sterling Deans. Architects have a role to play, together with

Gustav Lindenthal, celebrated bridge builder, was born in 1850 in Moravia. There is endless confusion about his formal education, which may have ended early. Claims to university degrees, for example, may have been confused with honorary degrees granted much later. With limited formal training, Lindenthal spent some time working for the Swiss National Railroad before departing for America in the mid-1870s—where self-educated engineers were readily accepted.

In Philadelphia he worked on the erection of major display buildings for the 1876 Centennial in Fairmount Park before moving on to Pittsburgh where he designed and built both the Smithfield and Seventh Street bridges, spans noteworthy for using eye-bars (European influence) rather than cables. From the Iron City he moved to New York City, where the need for bridge construction was great and he was unknown.

In Gotham, Lindenthal generated bombastic plans for a Mid-Hudson Bridge at Poughkeepsie that were never realized. In the early 1900s he served as New York's Bridge Commissioner, and got deeply involved in the controversial Manhattan Bridge project, again advocating eye-bars over cables. A change in the mayor's office removed Lindenthal from the Manhattan Bridge work and turned him to his next great passion: designing the Hell Gate Bridge. Lindenthal was hired by the Pennsylvania Railroad and charged with building this bridge to link the tracks of the Pennsylvania Railroad with the New York, New Haven and Hartford Railroad, creating a direct line from New Jersey through New York City to New England. The imposing four-track railroad bridge over the East River, with a viaduct over three miles long, was completed in 1916 and was the longest steel-arch bridge in the world at the time.

Gustav Lindenthal is well known to bridge historians and those familiar with the history of civil engineering in America. More recently his career has been celebrated by a 95-page chapter in Henry Petroski's Engineers of Dreams.

construction companies and erection crews. After the turn of the century, however, many structural iron workers, particularly in a place like New York City, belonged to the International Association of Bridge and Structural Iron Workers, which conducted the infamous "dynamite strike" in the first decade of the twentieth century.[3]

In the world we actually live in, bridge building is a process that extends far beyond designers, engineers, and erectors—to machine politicians, reform politicians (each with their own separate agendas), competing designers, competing bridge commissioners, zealous merchants, irate citizens groups and squabbling art commissions. While these webs of intrigue are amusing to read about and make for grand drama, the serious fact remains that all of this becomes dreadfully time-consuming and therefore brutally expensive. The financial burden that was added by this "human dimension" in bridge building will be underscored in the following pages.

Let the story begin with the distinguished Austrian-born bridge designer Gustav Lindenthal serving as bridge commissioner to "reform" Mayor Seth Low in 1902 and drawing up plans for a 1470-foot suspension bridge that would connect New York's Lower East Side to Brooklyn. Located less than a mile upstream from the Brooklyn Bridge, this new span was guaranteed to relieve congestion on the New York shrine.

The Manhattan Bridge was to feature eye-bar chains instead of the famous steel-wire cables for which the Roebling operation had become famous, and which might have been provided by the Roebling wire works.[4] Almost inadvertently, the new span was referred to as the "third East River Bridge" or the "Manhattan Bridge," and while both labels seemed commonplace and bland and therefore the object of criticism, somehow "Manhattan Bridge" became the tag that was never removed.

THE DEBATE OVER EYE-BARS

Eye-bars, as employed in the early 1900s, were elongated pieces of nickel steel, most likely ten feet in length, with "eyes" or holes in each rounded end. They were connected to one another by inserting a steel pin through the overlapping eyes of two separate bars. When this process is repeated enough times, it produces a chain that can be used to do the same job that *Roebling*'s wire cable did for the Brooklyn and Williamsburg suspension bridges. While eye-bars had not been terribly popular in America, Lindenthal had used them on the 7th Street Bridge in Pittsburgh in the 1880s and had become a staunch advocate of the concept. Eye-bars had also been employed in Budapest (a setting that lacked the capability to manufacture wire cable and possibly was too nationalistic to import it), and in Great Britain. Accordingly, they should not have been viewed as some exotic rarity, but somehow were so in the United States.[5]

To be sure, the suspension bridge could have been built with either wire cable or eye-bar chains. Yet, the debate that ensued between 1902 and 1905 was extraordinary, and it was reported with great consistency in *Engineering News*, *Scientific American*, and *The New York Times*.[6] One system would be stronger than the other; one would be more expensive; one would outlast the other; one would be more aesthetically pleasing than the other. The debate raged on. At one point the editor of *Engineering News* felt it necessary to remind readers that the preponderance of letters favoring wire cable that he was publishing should not be misconstrued as an official endorsement by the journal.[7] While most of the debaters were civil engineers who frequently permitted their calculations to speak for them, personal agendas could never be ruled out.

Gustav Lindenthal, for example, found that his commitment to eye-bars had been reinforced as a result of a dispute with the Roeblings just a few years earlier. Lindenthal had ordered wire cable from the Roeblings for the Williamsburg Bridge, and the Trenton firm had not delivered as rapidly as he would have liked. Lindenthal was miffed, and believed he was entirely within his rights to be so. From the Roeblings' perspective, however, they had done everything in their power to meet the contract on time. Indeed, they had started construction of a new five-story plant that would employ 300 workers to expand their production capacity—it simply had not been done rapidly enough to satisfy Lindenthal.[8]

On December 31, 1903, the editor of *Engineering News*, conceivably with a sense of exasperation, argued that perhaps the entire project was unnecessary. After all, where New York once had bridges and tunnels to convey trolleys, subways, and elevated lines on only four tracks, the city now had ten—and soon would have twenty-two. Perhaps New York City should wait five or ten years until commuters really needed the Manhattan Bridge and find the necessary $15,000,000 at that time.[9]

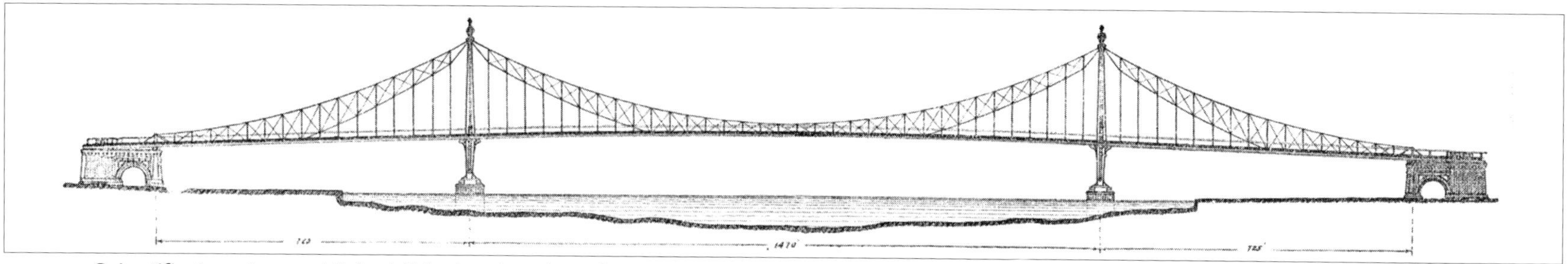

Scientific American published this drawing, based on the elevation plan of the Department of Bridges, in January 1904 under the caption, "THE NEW MANHATTAN SUSPENSION BRIDGE ABOUT TO BE BUILT ACROSS THE EAST RIVER, NEW YORK."

SHIFTING POLITICAL SANDS

The tempest over eye-bars versus wire cable could go on for several years only because design and construction had been put on hold. The election of 1903 returned control of the mayor's office to George B. McClellan, Jr., and the Tammany machine on January 1, 1904. This meant that "reformer" Seth Low was out—and so was his bridge commissioner, Gustav Lindenthal, and his design. The foundation piers that had already been constructed (starting in 1900 at a cost of $1,500,000) were "dead in the water," so to speak, and no structural steel would be raised in the immediate future.[10]

A delay of three or four years might have seemed like an eternity to the "strap-hangers"—people caught in the rush-hour crush on the Brooklyn Bridge—or to merchants who were sure this was costing them money in lost sales. How can we explain this? On the one hand, the project should not move forward until the engineering dilemmas had been resolved. More realistically, however, a new administration meant a transition—and a predisposition to put aside the work that was done before. Thus, Lindenthal's plan was cast aside (the other guys are never any good), and the new bridge commissioner, George E. Best, instructed his new chief engineer, O.F. Nichols, to draw up new plans—which he did with the aid of architects Carrere and Hastings. The new design itself, adding $100,000 to $280,000 to the cost of the project, not surprisingly, substituted wire cable for eye-bars.[11] Earlier, the Board of Aldermen had garnered 41 votes to defeat Lindenthal's $6,533,000 eye-bar appropriation.[12] Phoenix Bridge received contracts for the superstructure between August 27, 1906, and January 5, 1910; most of the erection took place in 1909.

A story in *Scientific American* for May 6, 1905, entitled "The Manhattan Bridge Scandal" ripped the new administration for being so slow to put out bids, condemn property, level homes in the path, and start work on either the anchorages or the towers. The article further pointed out that the new bridge commissioner had no credentials that justified overruling Lindenthal, and that the ineptness of the new regime would add $2,000,000 to the cost of the project![13]

Looking at this politically, it seemed that nothing related to the Manhattan Bridge project could possibly move forward until Tammany was assured that many of their faithful would be hired for at least construction of the portals, anchorages, and roadways leading in to the bridge if not for the more dangerous jobs associated with the erection of structural steel. Whatever the reason, no steel was hoisted skyward prior to 1906.

DISASTER IN QUÉBEC, PANIC ON WALL STREET, 1907

For an engineering project that needed no more delays or wrenches in the gears, the second half of 1907 was more than a bit unsettling.

At the end of the work day on August 29, 1907, the south anchor arm of the Québec Bridge—a cantilevered bridge under construction—collapsed into the St. Lawrence River and killed 75 of the 83 men at work on the arm. Beyond being one of the great bridge disasters in history, the Québec collapse was also significant because it was a project of the Phoenix Bridge Company of Phoenixville, Pennsylvania—the very firm that was providing almost all of the steel superstructure for the Manhattan Bridge, with the exception of Roebling's wire cable.[14] Accordingly, if Phoenix Bridge couldn't be trusted to build safely and effectively over the St. Lawrence, how could they be trusted with the work over the East River? Indeed, with the prospect of a deluge of litigation and bankruptcy hanging over the Phoenix Bridge Company, it became necessary to redraw the contract so that the bridge company's parent firm, Phoenix Iron and Steel, became the party responsible for the superstructure.[15]

What's more, the general public and engineers alike would now demand some extraordinary care where safety and inspection for the East River project were concerned. At the same time, in this climate of fear and uncertainty, the parties responsible for the erection of the Queensboro or 59th Street Bridge took the Pennsylvania Steel Company of Steelton, Pennsylvania, to court to demand that new (and allegedly safer) specifications and standards be drawn up for their span. Interestingly, the court looked beyond the emotions of the moment and upheld Pennsylvania Steel and the original standards.[16]

Another dramatic and disconcerting piece of news came to New York's citizenry on October 22 when the Knickerbocker Trust Company closed their doors in an effort to stave off a run. This signaled the onset of the Panic of 1907. While there is no reason to recount this well-known story here, and there was no major linkage between this financial dislocation and the sale of bonds to finance the Manhattan Bridge (at least so far as the author is aware), the Panic still sent a chill wind through the financial markets and actually altered the contract between Phoenix Bridge and the Ryan-Parker Construction Company of New York City. Under normal circumstances Phoenix Bridge would have had the right to cancel their contract with Ryan-Parker for nonpayment. The new contract, however, stipulated that "if there should be a great stringency in the money market, resulting in a general suspension of payment by the bank, similar to that which occurred in 1893, then the right of the Bridge Company to declare the contract at an end shall be postponed for so long as the said suspension of payment shall continue...."[17]

OTHER SNARES AND SNAGS ALONG THE WAY

A host of engineering decisions were part of the process. For example, if the Brooklyn Bridge was such a grand achievement, why not emulate the structure by having masonry rather than steel towers? The answer was that masonry towers required a base twice as large, the construction time would be two to three times longer, and the cost perhaps as much as five times greater.[18] While financial considerations alone would have been conclusive, there was also the belief at the turn of the century that masonry towers were essentially medieval and therefore old-fashioned. Steel, by way of contrast, was the wonderfully new material—symbolic of the modern technological America.

On a more difficult matter, civil engineers in the early 1900s evidenced profound interest in deflection theory. Austrian civil engineer Josef Melan and Leon Moisseiff saw potential for applying this in bridge building, and J.A.L. Waddell showed how this might be

Leon Solomon Moisseiff was born in Latvia in 1872. He was a student at the Baltic Polytechnic Institute in Riga circa 1890, but left for the United States in 1891 without a degree. He resided in New York City, enrolled at Columbia, and took a civil-engineering degree in 1895.

Focusing his efforts on urban transportation, consciously or unconsciously, Moisseiff went to work for New York Rapid Transit and later the Bronx Department of Streets before being hired by the New York Department of Bridges in 1898. In this setting he became a colleague of Gustav Lindenthal and devoted considerable time and effort to the design and erection of the Williamsburg, Manhattan and Queensborough spans. By 1915 he became an independent consultant.

Moisseiff is best known in engineering circles as a pioneer in deflection theory, whose application produced modern bridges that were both slender and flexible. He contended, and proved with a mathematical model, that as the dead weight of a bridge grew, the stiffness of the deck could diminish.

One scholar suggests that Moisseiff likely played some consulting role on every large suspension-bridge undertaking in this country during the 1920s and 1930s. Henry Petroski argues that Moisseiff was the key designer of the Bronx-Whitestone, George Washington, Mackinac and Tacoma Narrows bridges—whether he received credit or not.

When Moisseiff died of a heart attack a few years after the Tacoma Narrows disaster of 1940, his colleagues were predisposed to celebrate his contributions rather than assign blame. Additional insights into Moisseiff's contribution to deflection theory can be found in Jameson W. Doig and David Billington's "Ammann's First Bridge: A Study in Engineering, Politics, and Entrepreneurial Behavior" in *Technology and Culture* (July 1994), Vol. 35, No. 3, 537–567.

done. The application, however, was in its infancy in the first decade of the twentieth century, and flourished only in the 1920s. Deflection theory argued, for example, that "the stiffness of the bridge's deck could decrease as the load increased. Since longer bridges tend to be heavier,... long suspension bridges could be very flexible."[19] Suspension bridges were now deemed far stronger than previously imagined: the curved cables added strength—according to the new mathematical model. In effect, newer suspension bridges could be lighter, stronger, and sway in the wind—at least until the Tacoma Narrows disaster. The net result was that far less steel was needed. This saved money, thereby rendering any span more politically viable.

Within the civil engineering fraternity, deflection theory constituted the clever or new or "sweet" solution. The only problem in the very early 1900s was that, practically speaking, it encouraged "underbuilding." Put another way, it might have been fine to erect the Manhattan Bridge with the aid of deflection theory, but history indicates that the structure has been in the process of being reinforced and upgraded—seemingly forever. Currently it is undergoing twenty years of intensive rehabilitation by Weidlinger Associates, Inc., a New York City-based civil engineering firm. Indeed, an insert on "Manhattan Bridge" in the Weidlinger Associates' newsletter for spring 1996 points out:

> The Manhattan Bridge has always been subject to large movements, resulting in massive cracking. Prior to Weidlinger ... rehabilitation work had been ineffectual ... The Manhattan was the first suspension bridge to be designed by the deflection theory,... At the time ... available technology did not allow for the theory to be applied in the detail required by the bridge's complex structural system."[20]

In brief, what was the *true* cost of a structure that required endless improvement and attention throughout its life?

For reasons conceivably known to no one at this time, a design based on deflection theory was combined with a decision to run the heaviest modes of transportation (subway and elevated cars) on the outer as opposed to the inner lanes of the bridge. The net result was an unusual amount of torque on the outermost members of the bridge—causing the members to suffer damage and require attention.

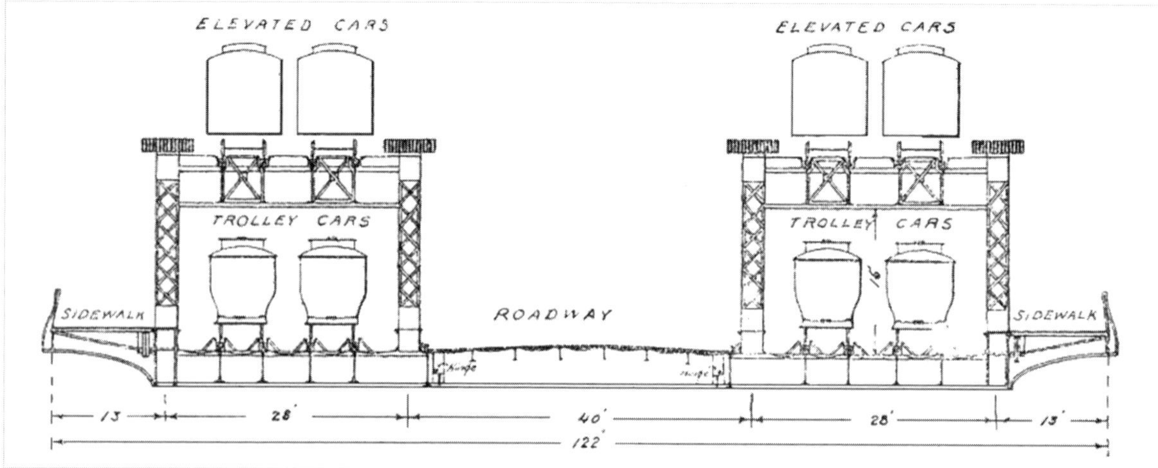

"SECTION THROUGH FLOOR AT CENTER OF SPAN" Drawing from Scientific American, January 23, 1904, showing the outside position of the tracks for subway and elevated cars.

A MAJOR OPPORTUNITY AT THE RIGHT TIME FOR PHOENIX BRIDGE

The aforementioned technical debates and political barriers delayed the letting of contracts for the superstructure to Phoenix Bridge. This delay, however, proved beneficial to the bridge company as the contracts were let between August 27, 1906, and January 5, 1910.[21] In essence, it allowed the firm to be engaged in one of the more prestigious projects they would ever be part of—at precisely the time that the Québec disaster was calling their competency and reputation into question.

Phoenix Bridge had contracts to provide the structural steel for anchorages, towers, trusses, etc.—essentially the lion's share of the superstructure. *The New York Times* for February 28, 1909, reported a shipment of steel to Bayonne on sixteen flat cars priced at $3,578,316, and estimated Phoenix Bridge's profits at $424,544 or roughly 11.9 percent.[22] On the Manhattan Bridge project, however, Phoenix seriously underestimated the cost of labor on the anchorages and trusses, and ended up with a profit of only $312,279 or 8.7 percent.[23] Not a great outcome for a firm that expected a double-digit rate of return, but, again, it was a great public relations counterweight to the Québec episode.

The Phoenix Iron Company works in Phoenixville, Pennsylvania, in 1906 (above), with the Schuylkill Canal in the foreground. Below: the Phoenix Bridge Company yard. (Hagley Museum and Library)

WHAT ECONOMIC DIFFERENCE DID IT ALL MAKE?

The preceding details endeavor to demonstrate what many well-informed people already know: namely, that projects that might readily be described as scientific or technical are often drowning in a sea of human controversy. As members of the Society for the History of Technology are always anxious to point out, bridge building (or any other grand technological effort) is a very human process. This matters to the citizens and bond holders because it renders already expensive jobs considerably more so.

The Queensboro and Manhattan bridges were both completed in 1909–1910. In some respects they were very different: the former was cantilevered, decidedly more ornate (it served a better part of town), had a total length of 8,600 feet (two channel spans of 1,182 and 984 feet; one island span of 630 feet), a width of 88 feet, and 11 lanes for vehicular traffic. It was built for about $20,000,000. The latter was a suspension bridge, with a fascinating portal, colonnade and plaza, a total length of 6,855 feet (main span of 1,470 feet), a width of 120 feet, and 11 lanes for vehicular traffic. It was built in the same town at the same time for $26,000,000—roughly 30 percent more.[24] Arguably, the Manhattan Bridge (in the "wrong" part of town) was something of a political orphan while the Queensboro or 59th Street Bridge was supported by a muscular uptown merchants association and a citizenry that would never have tolerated comparable delays.

The Manhattan Bridge was credited with relieving congestion on the Brooklyn Bridge, as anticipated. *The New York Times* for January 30, 1910, reported that on January 26, 1909, some 4,407 vehicles had crossed the Brooklyn Bridge while "yesterday's results showed that only 3,263 vehicles passed over the bridge."[25] This constituted a reduction of roughly 25 percent.

Time is money, delays are expensive. Perhaps we will never know what the Lindenthal design with eye-bars would have really cost, but it's hard to believe that it would have been more expensive than the meandering path that was followed! One of the prime historical complaints against the urban political machine was that it drove up the cost of building the urban infrastructure. Obviously.

Above left and left: segments for the towers of the Manhattan bridge being assembled in the Phoenix Bridge Company's fabricating shop.

Above: Canal boats, including one alongside the dock, are being used as construction vehicles in this photograph of the towers on March 2, 1908.

(Phoenix Bridge Collection, Hagley Museum and Library)

The high steel workers' view of Manhattan and the East River, photographed from the top of the Brooklyn tower on August 3, 1908. The walkways, from which the men wrapped the cables, are seen clearly here. (Collection of The New-York Historical Society)

chapter three

THE HIGH STEEL WORKER'S WORLD

Two things are abundantly clear about the hoisting and erection of the Manhattan Bridge superstructure: the work was brutally dangerous; and, perhaps because it was, high steel workers had their own distinctive subculture—a subculture predictably awash in stories and tradition. Officially, this was a union job (with a thirty-foot banner announcing same) and thus the workers were members of the International Association of Bridge and Structural Iron Workers (IAB&SIW). This organization was totally new for bridgemen, who had been unrepresented throughout the rise of the union movement in the nineteenth century. As rugged individualists who roamed America from one job site to another, they were a challenge to organize. At the same time, by 1909 roughly 10,000 men paid dues to the IAB&SIW.[1] In return, the union provided some political muscle, occasional legal representation, a death benefit of $100, injury benefits, and *The Bridgemen's Magazine*, which was issued monthly.

The magazine, predictably, voiced its strident opposition to the National Erectors' Association's advocacy of the open shop movement, but then filled the rest of the pages with homilies, family photos, poetry, death notices, and a variety of advertisements for boots, gloves, and work clothing. Jewelers' advertisements for diamonds suggest the high steel men were well paid, and they were, earning roughly $5 to $5.50 per day or $25 to $27 per 5½-day week at a time when many industrial workers earned only $2 to $2.50 per day.[2] This high compensation paid to bridgemen, connectors, derrickmen, planers, riggers, and riveters was really a function of the risk they took as opposed to the skills they possessed. Readers might have logically concluded that the union seemed intent on being something of a civilizing, stabilizing, and (literally) sobering influence on a notoriously wild and hard-living cadre of men. The IAB&SIW under the leadership of John J. McNamara of Cleveland, however, was not consistently interested in altering or tempering the iron worker's "roughneck" image, particularly because it could be used indirectly as a negotiating weapon. Indeed, the union conducted the now-famous "dynamiting strike" in the early 1900s that reached its crescendo on October 1, 1910, with the use of nitroglycerin on the *Los Angeles Times* building, killing twenty and doing over $500,000 in property damage.[3] In brief, the coming of the union, together with strikes utilizing calculated and calibrated violence, shaped and reshaped bridgemen. This does not mean that some observers could not choose to depict these daring men as mythic, storybook figures.

For example, in 1908 Ernest Poole endeavored in his own literary way to capture the essence and ethos of high steel workers in "Cowboys of the Skies," an eleven-page article published in *Everybody's Magazine*, featuring photographs by Arthur Hewitt and Frederic Colburn Clarke.[4] This flattering portrait, not surprisingly, was reprinted in *The Bridgemen's Magazine* in February of 1909.[5]

Poole divided structural iron workers into two groups, those who worked on skyscrapers and those who worked on bridges, pointing out that men generally did not shift their work preference from one to the other. More critically, the two groups had so much in common that any other distinction would have been meaningless. Poole described the stereotypical figure as one who "wore brown canvas trousers tucked into his boots, a grimy jumper, a shirt wide open at the throat, buckskin gloves frayed by hard use, and an old slouch hat.... His lean tanned face was set in a puzzled scowl...." [6] The author attempted to capture the sense of exhilaration and ecstacy that cowboys of the skies experienced: "Up here the wind blew fresh and clean and the details of life dropped off into space, and above me on the open steel ... bristled up into the heavens some two hundred grimy men.... Silent men in the roar of the steel...." [7] Poole's workers were romantic innocents who lived to take risks and breathe the air at intoxicating heights.

Who, precisely, were these men and where did they come from? There is consensus that the single greatest source was seaman or mariner—someone who had spent many years climbing riggings and ropes and had conquered any fear of heights. The national origins, as described by Poole, are almost amusing: "Americans, English, Irish, French-Canadians, Swedes, now and then an Italian. And in the New York gangs this year two full-blooded Indians ... cool-headed and sure, a stolid pair who have little to say, climbing about on the dizzy heights...."[8]

The likelihood of only two Native Americans working in the "New York gangs" or on the Manhattan Bridge in 1908 seems rather remote as the New York membership of the IAB&SIW that year was 1,761, with an additional 550 members in Brooklyn.[9] That there were significant numbers of Native Americans in this occupation is evidenced by *The Bridgemen's Magazine* for October of 1907, which reported that at least 33 of the 75 workers, or 44 percent, killed in the Québec Bridge disaster of August 1907 were "Indians."[10] Yet another source reported the presence of "a lot of Indians" working on the Hell Gate Bridge, built between 1912 and 1917.[11]

INJURY AND DEATH

While there is agreement that erection crews do dangerous work, the more pointed question becomes: how dangerous? Was there anything comparable to the contemporaneous Québec Bridge episode on the Manhattan Bridge site? The quick answer is no. But, then, what does injury and death look like on more "normal" or "representative" projects? To find more precise answers, *The Bridgemen's Magazine* provides an accounting for the union, and *The New York Times* notes individual accidents on the Manhattan Bridge.

Bridgemen's for October of 1908 reported that seven percent of their workers were injured each week, and that the average disability time was roughly seven weeks—crude calculations point to the loss of half the work force in less than two months. If these data strain credibility, and possibly they should, it helps to remember that a labor magazine dealt in advocacy. Objective scientific truth may have been a luxury they believed they could not afford. Average disability time of seven weeks suggests the injuries reported were something more than minor bruises or sprained ankles. The IAB&SIW claimed they paid out $5.00 per week for two weeks[12] in disability benefits—at a time when the concept of workmen's compensation was being debated in state legislatures across America, but generally not implemented much before The Great War. There is really no good reason to contend that the iron workers were guilty of what Frederick Winslow Taylor would have

View from the Manhattan side over the East River to Brooklyn on December 8, 1908, before the hangers were installed to suspend the roadway. Docks lined the river on both sides.

Opposite: Anchorage and cable installation on the Manhattan side on August 3, 1908, and the view from Brooklyn on February 11, 1909, after the suspenders or hangers had been put in place.

These photos from the collection of The New-York Historical Society show dramatically how dangerous the work could be.

called "soldiering" or faking the injuries. On the contrary, the worker's pride in his own mental toughness probably motivated many of the injured to work despite their injuries.

Over time, insurance companies became more aggressive in their demands for innovations to promote greater safety. The suspended falsework on the Manhattan Bridge featured four platforms eight feet wide, each with two wire-rope hand rails, and sheave towers for hoisting materials. "The platforms are braced against excessive deflection and against wind pressure by back stays from each tower and by storm cables curved upward and connected to them at intervals of about 54 feet by suspender ropes." A few years later on the Hell Gate Bridge work site, hand ropes and temporary wooden railings were utilized as well as steel cages with platforms.[13] Alas, these devices never eradicated the great danger that persisted in structural steel work. A National Safety Council report for 1931 revealed that bridgemen still performed work roughly twice as dangerous as either more general construction work or coal mining.

The New York Times reported four noteworthy accidents stemming from the erection of the Manhattan Bridge superstructure, all taking place between January and August of 1909. On January 10 two were injured, one seriously, when a guy rope broke and a derrick fell across East Broadway. Edward Haggerty, the foreman, was arrested and charged with assault.[14]

On April 13 the paper reported a death on the span on a "Blue Monday." (Iron workers chose to believe that bad things happened on Mondays. They also chose to believe that accidents occurred in groups of three.) In this spectacular episode, John McShane was hit with a blast of compressed air from a pipe and hurled 120 feet to his death.[15]

On July 17 *The New York Times* told the tale of what happened when a defective plank gave way. Basically, a cracked board broke and sent three bridge workers plummeting 135 feet into the East River. Remarkably, only one of the three died.[16]

The final accident, cited in the paper for August 20, suggests that the bridge work frequently attracted a healthy number of "sidewalk supervisors." In this instance a twenty-foot wooden shield went crashing into a crowd of onlookers, injuring eight, one of them mortally. Foreman George Kasper was arrested and charged with criminal negligence.[17]

The fact that two of these episodes, both coincidently involving the general public, were followed by arrests and the filing of criminal charges seems a bit puzzling. (Consulting an attorney shed no additional light on the matter—though an explanation may well be buried somewhere in common law.) Apparently, either the foremen were so patently negligent in their structuring and supervision of the work site that criminal charges were warranted, or the arrests were a convenient public relations ploy designed to provide the newspapers and the public with a sense of immediate justice.

What about "death rates"? A careful gleaning of the death notices (including cause of death) in *The Bridgemen's Magazine* for twelve months in 1909 revealed 136 listings. Of this number, 75 or 55.1 percent were rather clearly "job related." The most common cause of death among the job related was "fractured skull" while the most common among other causes was "tuberculosis" or "consumption."[18] Calculating 75 fatalities as a percentage of the union membership of approximately 10,000 indicates that the death rate generated in this crude fashion was three quarters of one percent.

Luke Grant, reporting on behalf of the United States Commission on Industrial Relations in 1915, used data for the fiscal year 1911–1912 from the IAB&SIW to demonstrate that the union paid out 124 claims that year and that 109 of these were a result of accidents on the job. The 109 out of a membership of 10,928 constituted a death rate of one percent of the membership per year.[19]

The United States Bureau of Labor Statistics derived a figure for fatalities of 1.2 percent for the early 1910s.[20] Thus the truth most likely resides somewhere between three quarters of one percent and one and two-tenths percent per year. In this instance the concept of one percent—normally thought of as a very small number—is actually very large.

WHO SHOULD BE BLAMED?

While it is true that the Phoenix Bridge Company and Ryan-Parker Construction Company of New York City maintained a convenient truce with the IAB&SIW union during the erection of the Manhattan Bridge, this did not erase a history of animosity and tension that sharply influenced any debate over blame and guilt when workers were injured or killed on the job. For example, *The Bridgemen's Magazine* for October of 1907 reminded the membership that the Phoenix Bridge Company had a "bad reputation since the Louisville wreck" of 1893 in which twenty were killed when a span under construction was, according to company attorneys, destroyed by a cyclone.[21] In that episode, representatives of the bridge company, rather experienced in what today would be called "damage control," quickly contacted the widows and relatives of the deceased in the Louisville area and offered them a one-time cash settlement of $300. Apparently, most or all of these families signed releases in return for the cash. In 1907, in the wake of the Québec Bridge collapse, the union tried to warn families against signing any papers presented by agents of the legally liable Phoenix Bridge Company, which had kept eighty-three workers busy on the ill-fated south arm even after structural flaws had been identified.[22] The IAB&SIW contended that they had already allocated $500 for attorneys' fees and would be filing at least twelve damage suits.[23]

By the early 1900s the men from Phoenixville were proficient in responding to accidents and limiting corporate liability. Indeed, a piece of correspondence in the company pa-

Rivet barrels stored behind chainlink fencing, seen from Water Street at Mechanic's Alley. (MCNY)

pers basically advised their agents to get to the scene early, pay doctor and hospital bills, appear generous, get the widow to sign a release in return for $200, and get out of there! [24] Given the union's experience in Louisville and Québec, there was no reservoir of trust to draw on between the union and the firm that provided the steel for Manhattan Bridge.

By way of contrast, the companies were free to look at the workers' culture and find plenty of cause for complaint or blame there. Poole's portrait of the high steel workers cited earlier in "Cowboys of the Skies" cuts two ways. The endless daring feats of riveters that Poole romanticized looked very different to an insurance company. For example, a worker hitching a ride (when the foreman was not looking) on a piece of steel being hoisted may have enjoyed the ride—particularly if he was not knocked off by a gust of wind. A bridgeman jumping from one piece of steel to another was having a grand time—if he did not lose his balance. A worker 150 feet in the air maintaining his balance by leaning into the wind was doing something very clever—until the wind stopped. A bridgeman drinking on the job site may have lost some of his inhibitions—but that was a notoriously bad place to do so. Clearly, the cultural values of the high steel worker increased the danger on the work site, and this was never completely eliminated. Not by unions. Not by insurance companies. Not by government reports. Not by safety devices.

To be sure, the question of blame is rarely resolved to everyone's satisfaction. The union remained steadfastly convinced, with some justification, that workers would never get justice from the courts or a Canadian Royal Commission inquiry. *The Bridgemen's Maga-*

zine sneered at the contention that the Québec disaster resulted from a "design flaw." The IAB&SIW viewed this as simply a clever way to deflect legal liability away from any individual or any firm.[25] In short, high steel workers on the Manhattan Bridge led precarious lives; and, furthermore, believed they had very little recourse in the face of death or injury—regardless of who was at fault.

THE MINE FIELD OF LABOR RELATIONS

The uneasy truce between Phoenix Bridge and Ryan-Parker on the one hand and the IAB&SIW on the other was remarkable in light of the broader hostilities and labor disputes that existed between the National Erectors' Association and the iron workers' union. The National Erectors' Association of the early 1900s, under the leadership of Walter Drew of Grand Rapids, Michigan, was one of the most single-minded organizations imaginable. Comprised of most of the firms that carried out the erection of steel bridges or other structural steel work, and predictably led by the American Bridge Company (a subsidiary of United States Steel), the organization existed solely to promote the "open shop." With the aid of Drew's aggressive tactics the NEA hoped not only to secure existing open shops, but to roll back the union where plants had been organized.[26] The stage was set for extraordinary conflict.

In the late summer of 1905, John J. McNamara and the IAB&SIW launched a strike against American Bridge, headquartered in Ambridge, Pennsylvania, for higher wages. By early 1906 the strike had reached New York City and during the winter months New York employers responded by recruiting a legion of "outsiders" or strikebreakers, possibly equal in number to the few thousand who had walked out. In the face of this pressure the IAB&SIW both sought and found a temporary compromise in order that union jobs might be saved. Above all else, the union understood its own vulnerability. Beyond the normal disparity of wealth and power that separated labor and management, the IAB&SIW in New York was further overwhelmed by the reality of imported workers and an effective espionage campaign that undercut union planning and strategy.[27] On the surface, labor did not have a chance—or so it seemed.

THE PERSUASIVENESS OF VIOLENCE

The principles of military science concerning the conduct of warfare are very clear. When you cannot match your opponent in manpower, resources, or firepower, you resort to stealth and creative new strategies. In essence, you resort to guerrilla warfare. The IAB&SIW understood this principle. Accordingly, rather than accept defeat, union workers turned to what they euphemistically referred to as "entertainment committees," or "slugging gangs," and dynamitings—perhaps with the understanding that nitroglycerin was substitutable for dynamite. The union took great pride in identifying "sluggers" among their own rank and file as opposed to having to go outside to hire "specialists" as some other unions did.[28]

History records that the IAB&SIW resorted not only to violence, but to creative new forms of violence in their dispute with the NEA in an effort to "level the playing field." Stories of strikebreakers being beaten in New Orleans, Cleveland, and New York, and some of these beatings proving fatal, represented nothing new in America's long and violent labor history. The dynamitings, however, were a departure. The explosions at bridge sites, beginning about 1905–1906, were roughly contemporaneous with the Western Federation of Miners' use of dynamite in labor disputes in western mining regions.[29]

One of the distinguishing marks of the IAB&SIW dynamiting campaign was the absence of injury to people. Consistently, from 1905 through October 1, 1910, the targets were property, namely bridges under construction and stockpiles of structural steel at the job site. In many respects this was a brilliant new strategy in that it reduced the NEA's advantage and generated lots of publicity without endangering human life. Not surprisingly, the Phoenix Bridge Company was targeted by the union. Under the leadership of David Reeves, the firm may have been the most extreme member of the National Erectors' Association, taking an anti-union position even beyond that of other NEA members.[30]

The New York Times for April 27, 1908, reported "Dynamite Damages Unfinished Fall River Structure at Least $15,000." Phoenix Bridge had a $750,000 contract for a bridge over the Taunton River in Bristol County, Massachusetts, that would serve to better link the towns of Fall River and Somerset. The span was about two months from completion when the explosion occurred. Police determined that the perpetrators had used a boat and planted fifty pounds of dynamite on a granite pier nearest the Fall River shore, which they fired by a long fuse. The brief article concluded that if the foundations were damaged, the estimate of $15,000 would be too low.[31] There was no mention of any connection between the explosion and labor difficulties involving the IAB&SIW and the NEA or Phoenix Bridge. Professor Sidney Fine in "The National Erectors' Association and the Dynamiters" in *Labor History* (winter 1991 issue) noted a relatively minor episode at the same bridge—if "minor dynamiting" is not an oxymoron. Fine mentioned the arrest of one George O'Donnell in the wake of the dynamiting of a pile of structural steel stored along the shore,[32] no doubt part of the effort to complete repairing the damage done by the April explosion. Thus, while the American Bridge Company was the primary target of the union, it seems clear that Phoenix Bridge had also won a place in their hearts as well. (Surviving records of the Phoenix Bridge Company make no mention of either incident.)

LABOR TRANQUILITY ON THE MANHATTAN BRIDGE ERECTION SITE

If the IAB&SIW and the members of the NEA were generally at each other's throats between 1905 and 1910, and if the erection of the structural steel for the Manhattan Bridge took place primarily between 1906 and 1909, then it stands to reason that the job site in question should have been a battleground. But it was not. It is also fascinating to note that the union deemed Phoenix Bridge to be "unfair" in Pittsburgh, unfair in Providence, unfair in Fall River, and unfair throughout the South, and yet never formally added the company to its official list of unfair firms published every month in *The Bridgemen's Magazine*.[33]

The company appears to have been a thorn in the side of the union just about everywhere in America, but somehow Phoenix Bridge was acceptable in New York City even as they demanded exemption from the city's eight-hour law for their contractor (Ryan-Parker Construction) as a condition of doing business.[34] How can this be explained? Was the IAB&SIW guilty of "selling out"? Phoenix Bridge records, unfortunately, shed no light on this matter. An amusing letter to the editor of *The Bridgemen's Magazine* of April 1908 claimed that the union members were performing admirably on the job. Brother Walter Godfrey "... knows what a good day's work is, and keeps scaffold movers, fitters up, riveters and all moving along. The work is being pushed to its full force. Brother Jim Davis ... makes some of the "roughnecks" step lively."[35] Yet, even if it were prudent to believe

Construction of the roadway extending from the Brooklyn Tower, as seen from Main Street, Brooklyn, on March 11, 1909. (Collection of The New-York Historical Society)

Stiff-legged derricks are visible as construction of the roadway proceeds. This view looks toward the northwest from the Brookyn side on April 9, 1909. A freighter is moving upstream approaching the bridge, and a ferry is in midstream on the far left. (Collection of The New-York Historical Society)

the union's praise of its own virtue, this should not be expected to explain labor peace on the job site. What was going on here?

Without perfect knowledge of what accounts for this accommodation, it seems reasonable to argue Manhattan Bridge constituted a very high-profile project, and that New York City was a special case. Furthermore, while the NEA and Phoenix Bridge may have generally had the upper hand in dealing with the union, they may not have wanted to risk a showdown in the heart of the union's strength. Labor trouble obviously would have disrupted erection on a project that was already destined to take a decade to complete. The citizenry of Gotham, many of whom were commuters, considered any further delay both scandalous and unconscionable.

An extensive body of correspondence between John Sterling Deans, chief engineer of the Phoenix Bridge Company, and his assistant Peter Szlapka on the one hand, and the Ryan-Parker Construction Company of New York City on the other, reveals the extreme anxiousness of the bridge company to produce the structural shapes required, pressure the Pennsylvania and Reading railroads to move the bridge steel to the Bayonne storage facility as rapidly as possible (not waiting till they had five carloads), and transfer the structural shapes by barge to the job site on the banks of the East River where Ryan-Parker would have to take delivery and thereby be obligated to make payment to Phoenix Bridge. By contrast, labor trouble constituted good cause for Ryan-Parker to reject a ship-

ERECTING THE FLOOR SYSTEM, SHOWING THE CORDS WHICH WILL CARRY THE ROADWAYS.

DISMANTLING THE FOOT BRIDGE—AN ARDUOUS TASK AT A DISSY HEIGHT.

ment. They could simply have argued the project had been brought to a halt, and that there was no room to store the steel on-site. In essence, Ryan-Parker played a "cat and mouse game" with Phoenix Bridge. The construction company might demand the shipment of structural steel on Monday and reject delivery on Friday. Furthermore, they could rarely be bothered to make payment.[36]

The Bridge Department of the City of New York was another major source of delay in that they had to approve any changes. These city workers functioned only at their own pace and, predictably, resented being pushed. John Sterling Deans wrote to Ryan-Parker on January 28, 1907, seeking to determine the relative configuration of roadways, trolley tracks, and subway lines on the new span—having been led to believe that changes were in the process of being made.[37] Almost two months later, on March 15, 1907, Deans wrote essentially the same letter to Ryan-Parker, reiterating we "desperately need instructions on the configuration of roadways on the bridge."[38] It is unlikely that Phoenix's chief engineer had any idea in 1907 just how important that decision on configuration would turn out to be.

Another reason why Phoenix Bridge worked with a sense of urgency, and therefore needed union cooperation, was they were smarting under criticism generated by the Québec Bridge collapse of August 1907. The best way to defend themselves may not have been with words or high-powered attorneys, but by completing a really major project for the rest of the engineering profession and public to behold!

The foregoing paragraphs begin to explain why Phoenix Bridge and the NEA saw peaceful coexistence as the wisest course of action. Bridgemen would still walk off the job when a fellow worker lost his life in an accident, but basic agreement between labor and management prevailed.

Opposite: Construction of the Manhattan Bridge, a massive public project, was followed by city newspapers in order to satisfy the curiosity of their readers. These photos are from the The Saturday Evening Mail on March 27, 1909. (MCNY)

CONCLUSION

While the building of any great bridge ought to be the basis for exciting history or a darn good yarn, the story of Manhattan Bridge was enhanced by the excitement surrounding the IAB&SIW's dynamiting strike, an episode that wreaked havoc on the work of the Phoenix Bridge Company at Fall River, Massachusetts, but not on their work a hundred miles away that spanned the East River from Canal Street to Flatbush Avenue. Between building the piers and anchorages on the one hand, and completing the hoisting and riveting of structural steel on the other, the project consumed a decade. The cowboys of the skies played their starring roles between 1906 and 1909 and generated nothing more than anticipated instances of deaths and injuries—surely nothing comparable to the horror story of which Phoenix Bridge became a part on the St. Lawrence River in Québec. The bridgemen tossed and caught white-hot rivets to the delight of onlookers, and earned good wages for their five-and-a-half-day weeks. When it was all over, they were conscious of having been part of one of the great engineering projects of the twentieth century.

chapter four

THE LOWER EAST SIDE AND SOUTH BROOKLYN LINKED

The First Forty Years

If the life expectancy of a modern iron, steel, and nickel-steel suspension bridge were forty years, then the Manhattan Bridge could have been deemed a success. A demolition crew could have been hired in the early 1950s to dismantle the span that, theoretically, could have been replaced by a few new tunnels (this almost happened). End of story.

Alas, designers, engineers, government officials, and the citizenry have come to believe, with some justification, that these metallic behemoths should serve the public for at least a century—after all, the iron bridge over the Severn Gorge in England has been in place since 1779! Where the Manhattan Bridge is concerned, only the first four decades were generally positive. After 1953 it was patch, patch, patch.

The addition of the Williamsburg, Manhattan, and Queensboro bridges over the East River in the first decade of the twentieth century brought a number of changes to the New York area, many of which, with hindsight, now seem almost too obvious to mention. Hundreds of thousands of unmeltable ethnics moved from a brutally congested Lower Manhattan to Brooklyn, Queens, and Long Island. Many of these same people were destined to become commuters. Not surprisingly, land values east of Manhattan soared. None of this would have happened as it did apart from the new bridges and improved transportation in general.[1] Put another way, once-isolated neighborhoods now became part of the new metropolitan collage.

In particular, the Manhattan Bridge was expected to relieve congestion just a few blocks away on the Brooklyn Bridge, and data cited in an earlier chapter suggest this is

View of the Manhattan tower, showing the busy waterway below. A ferry is heading upriver on the far left of the photograph, and in the center of trhe river is a tug pushing a lighter to meet a steamship. The date is June 29, 1909, and the bridge is still under construction, with the high walkways still in place. The bridge was officially opened by Mayor McClellan on December 31, 1910, even though the four trolley and four subway tracks and the pedestrian walkway had not been completed. (Collection of The New-York Historical Society)

exactly what happened. The new span stayed out of the limelight during the first decade of its service, and in no serious way threatened to compete with the Brooklyn Bridge as an equally appealing edifice for lovers, jumpers, protestors, or artists. (Photographer Berenice Abbott did develop a passion for portraying Manhattan Bridge in the 1930s.) In many respects the new span was a modern and utilitarian agglomeration of structural steel destined to make no designer or engineer famous. Some even complained that the name was uncreative and mundane—though it is difficult to fathom why "Brooklyn Bridge" was a good name and "Manhattan Bridge" a bad name.

Strange as it may seem, the exact utilization of the various lanes and two levels of the span had to be determined over time. Actually, it became a process of evolution. For example, when the bridge first opened at the end of 1909, or shortly thereafter, subway tracks were added to the lower level and situated in the outside lanes. Their first use apparently came about sometime in 1912 by a trolley company with the unlikely name of The Manhattan Bridge Three Cent Line. The trolley line used the tracks between 1912 and 1915 until it got "kicked upstairs" to the upper level. Later, of course, motorized vehicles would drive the trolleys from the span altogether. Brian Cudahy, writing in *Under the Sidewalks of New York: The Story of the Greatest Subway System in the World*, reported that in June of 1915 a "two car BRT train ran tests on the Fourth Avenue line and across the Manhattan Bridge to a station under the Municipal Building on the Centre Street loop."[2]

The tests were successful, and Cudahy observed that on June 19, 1915, an eight-car ceremonial train jam-packed with public officials from New York and Albany, in addition to BRT officers, rode the rails from the municipal building in downtown Manhattan across the Manhattan Bridge and on to the West End Depot at Coney Island (Brooklyn) to celebrate the opening of the line.[3] Cudahy, a writer who loves the New York metropolitan subway systems, understands that the trains using the Manhattan Bridge might have been more efficient and rapid had they used a tunnel. He also comprehends the implications for bridge maintenance. At the same time, he fully appreciates how magnificent the view of the river, boats and barges, bridges framed by other bridges, and the New York skyline is for the commuter and day tripper.

Some of the best stories that surfaced relating to the Manhattan Bridge in its first decade of service concerned allegations, arrests, law suits, and dismissal of public officials stemming from the construction of the new suspension bridge, the sort of thing that the salty political observer Mr. Dooley would have labeled either "honest graft" or "dishonest graft." Sometimes the press simply made a passing reference. For example, in April of 1912 *The New York Times* observed that a certain Fosdick recommended that a man by the name of LaChiotti be dismissed from the bridge department for accepting a fee from the Ryan-Parker Construction Company for "professional services" rendered during the erection of the Manhattan Bridge.[4]

In a slightly more detailed account, politico C.J. Sullivan sued the Ryan-Parker Construction Company for $2,000,000 or "his share of the profits" based on the claim that Sullivan was personally responsible for Ryan-Parker getting the contract. Sometime in 1911 a jury awarded Sullivan the cryptic sum of $301,791, a figure that the New York Superior Court set aside in February of 1912 as "excessive."[5] Over seven years later *The New York Times* for July 3, 1919, reported that Timothy D. Sullivan, most likely a surviving heir, was awarded the grand sum of $322 in a final verdict.[6] Apparently the perceived market value of bid-rigging had suffered from deflation or serious shrinkage in an era of Progressive reform. In any event, during its first decade of service to Gotham, the Manhattan Bridge was simply a public asset contributing to the metropolitan area's growing transportation network. The primary beneficiaries were commuters who needed to cross the East River five or six days a week to get to work, or weekend "day trippers" headed for the beach, the boardwalk, and the thrills of one of America's premier amusement parks: Coney Island.[7]

THE 1920S

In July of 1922, in the face of growing traffic congestion, a degree of specialization was introduced in an experiment such that the Manhattan Bridge was closed to all horse-drawn traffic—now required to use the Brooklyn Bridge—reserving the much newer (by twenty-six years) Manhattan Bridge for motorized vehicles.[8] (Then as now, motorists grew impatient and angry when stuck behind slower-moving, horse-drawn vehicles, and engaged in "road rage" long before the term was invented.) Just a few days later the press reported the change was a great success, traffic was flowing beautifully, and letters of gratitude

The Brooklyn end of the Manhattan Bridge in the 1920s, showing double trolley tracks on each side. (MCNY)

were pouring in to the *Times*. In addition, overnight, the Manhattan Bridge became a good deal cleaner. Then, in a beautiful moment of naïveté, the reporter went on to say, "The up-

Berenice Abbott, who possessed an extraordinary passion for photographing Gotham, was born in Springfield, Ohio, in 1898. After a few years at Ohio State University she left in 1918 for Greenwich Village, where she did some acting, sculpting and writing. She became close to Dadaists Man Ray and Marcel Duchamp and joined their exodus to Paris in 1921 when the Village became a tourist attraction.

In Paris, Abbott became a fixture in the Montparnasse portrait studio of Man Ray, where she progressed rapidly and by the mid-1920s had earned a comparable reputation for her portrait work. She shifted to photography under the influence of Eugene Atget, a Frenchman who pioneered as a modernist and realist as he documented Paris between the 1880s and 1920s.

Early in 1929 Abbott returned to New York to publish a collection of Atget's work and was overwhelmed by the architectural developments in Manhattan during her absence. She rather quickly decided that she would do for New York what Atget had done for Paris: she would provide a visual record of skyscrapers, elevated lines, bridges, and other structures.

Initial efforts to generate funding produced not only failure but hostility; however, Abbott finally garnered support through teaching at the New School (1933 through 1958) and through the New Deal's WPA. Her intense love for the Manhattan Bridge is rather transparent in her work.

In 1997 Bonnie Yochelson edited *Berenice Abbott: Changing New York*, and in 1998 the Museum of the City of New York housed a "Berenice Abbott Retrospective." On April 17, 1998, the *Chronicle of Higher Education* further highlighted Abbott's legacy in an article entitled "Photography for a Scientific Age" that featured a magnificent view of the span from the pedestrian walkway.

per and lower roadways ... are now used to 100 percent capacity."[9] The perception of what constituted 100 percent capacity would change over time.

As F. Scott, Zelda, and friends poured back and forth across the East River bridges in the 1920s, a growing metropolitan population generated greater and greater traffic congestion. Between 1920 and 1930 the citizenry in the five boroughs grew from 5.6 to 6.9 million, an increase of roughly twenty-three percent.[10] By the spring of 1926, a regional planning commission was studying the transportation question in general and the tie-ups on the East River spans in particular.[11] As early as 1927 the New York Port Authority considered the need for a special bridge and tunnel authority, and during the first half of 1929 the citizen outcry for more bridges and tunnels became quite shrill.[12] One of the more creative responses to this outcry came from the administration of one of Tammany Hall's finest, the legendary Mayor James J. Walker (alias "our Jimmy" or the "night mayor"), who sprang into action and had the City of New York purchase both the franchise rights and the rolling stock of the trolley line operating on Manhattan Bridge. Only four days later the service was discontinued in favor of buses, providing a resounding popular victory for the all-powerful motorists of the area. The same motorists that resented slow-moving horse- drawn vehicles also resented trolleys along with their exclusive bridge lanes.[13]

THE DEPRESSION DECADE

The wretched economy of the 1930s did not rid our highways, roads, bridges, and tunnels of motor vehicles. Even though the number of registrations did decline nationally from 26.7 million in 1929 to 24.2 million in 1933, by 1940 the number had soared to 32.5 million; thus, cars, trucks, buses, and subway trains continued to pound the roadways and tracks of the span from above, and there were occasional subterranean assaults as well.[14] In early 1931, subway workers tunneling under the East River sought to protect the Brooklyn tower of the Manhattan Bridge. "A hole, three by seven feet, was made in each bridge foundation pier and an eleven-ton girder was inserted, with both ends resting on the cement columns."[15] Is there another city in the world where the foundation of a major bridge tower would have been subjected to a comparable threat or disturbance? The vignette illustrates the complexity associated with putting together a transportation network where

Manhattan Bridge in January 1940, after the modifications that were supposed to reduce the torsional flexing had been completed. This photograph, taken from the roof of 40 Monroe Street at the corner of Market Street on the Lower East Side (an area known as Knicherbocker Village), shows the intensively developed waterfront of the mid-twentieth century. Pier 30, a recreational pier, is the right foreground. (MCNY)

subway lines, tunnels, and underground pipelines and wiring systems of every conceivable variety criss-cross and twist around one another in vertiginous fashion.

The opening of the new upper express roadways on Manhattan Bridge—not all on precisely the same level—was a great excuse for a parade on June 18, 1931, led by Mayor Walker and borough presidents Hersterberg of Brooklyn and Levy of Manhattan. The march began on Bayard Street in the Bowery and crossed the bridge to Brooklyn, passing a reviewing stand at Johnson Street and Flatbush Avenue. The line of marchers included battalions of infantry from Governor's Island as well as sailors and marines from the Brooklyn Navy Yard. Ribbon-cutting ceremonies were held at each end of the bridge, the mayor delivered a speech before guests seated in the grandstand, and all of this was followed by a luncheon at the Hotel St. George.[16] Local politicos had found a reason to engage in a celebration of self. Everyone loves a bridge opening.

This improvement meant that the carrying capacity of the Manhattan Bridge increased from 65,000 cars daily to anywhere from 95,000 to 110,000 cars daily, depending on who made the estimate. By comparison, the Queensboro Bridge had a capacity of 92,000, but was planning to add capacity for another 30,000 to 35,000. At the lower end, the Williamsburg Bridge carried about 48,000 cars per day and the Brooklyn Bridge roughly 25,000.[17] To simplify the new traffic patterns, the upper roadway on the north side of the span would always be Manhattan-bound while the new upper roadway on the south side of the bridge would always be Brooklyn-bound, thereby avoiding the confusion that resulted from designating different traffic patterns during different hours of the day.

December 31, 1934, marked the twenty-fifth anniversary of the opening of Manhattan Bridge, and appropriate ceremonies were conducted on January 1, 1935, to celebrate the occasion.[18] Thus far there was no reason for the public to believe that this East River span was anything other than a well-designed and functional modern structure performing as anticipated. In August of 1935, when the Plant and Structures Department requested $4,000,000 from the city for re-paving East River bridge roadways, it must have seemed reasonable to assume that heavily traveled arteries normally required this.[19] By 1937–38, however, serious structural problems became public knowledge for the first time.

Early in 1937 bids were submitted to the Plant and Structures Department and a contract was awarded to the Taylor-Fighter Steel Construction Company of 570 7th Avenue for "steel alteration work" that would require roughly a year to finish. For $279,700 the contractor was going to complete alterations to "reduce side sway of the span under heavy traffic."[20] Unfortunately, this problem would never lend itself to simple correction. The weight of the subways trains, in particular, was problematic: each empty car weighed roughly forty tons. To this figure had to be added the weight of densely packed passengers, and then multiply the result by the number of cars in a given train—perhaps six, seven, or eight. Whatever a packed subway train weighed, it was surely a greater load than many spans were ever asked to bear. Day by day these trains rumbled incessantly across a Manhattan Bridge that sagged, swayed, groaned, and vibrated. What really seemed to matter, however, was that a few hundred thousand commuters per day got to work and got home again.

Suspension bridges, according to deflection theory, were expected to sag in response to heavy loads, and perhaps sway a bit; multiple trains on the bridge at the same time, however, were obviously never synchronized in their movement. This imbalance meant that beyond the seemingly harmless sagging and swaying there was also a great deal of twisting or torque. It was torque that proved so destructive to the structural steel on Manhattan Bridge in 1937–38, and it remained a major headache sixty years later in 1998 when Weidlinger Associates of New York City was custom-designing "torque tubes" for the span to solve a daunting problem that never stopped tearing at the structure.

A *New York Times* story for March 27, 1938, provided a few more details regarding the repair work to be done. Apparently it was time to install a new lower deck, and "... work on the supports carrying trusses at the towers will continue.... Pins, rockers, and other structural features of the bridge are being strengthened at a cost of $280,000. Subway trains using the bridge at night have been rerouted as a precaution." [21] On the eve of World War II a foreshadowing of some of the structural problems of Manhattan Bridge was in evidence.

Pedestrians walking across the Manhattan Bridge on November 11, 1936. (Berenice Abbott photograph, MCNY)

THE WAR YEARS

If bridges were generally deemed to be important in peacetime, they were seen as unusually important in wartime. If a foreign power, for example, had destroyed the East River bridges in 1942, this would have gone a long way toward turning America's largest city and financial capital into a partially staffed and therefore partially crippled metropolis. To illustrate the sensitivity to this problem, as early as July of 1939 in the debate over Battery Bridge versus Battery Tunnel, the argument was made that a bridge was far more vulnerable in wartime. Thus, the only prudent course of action was to build a tunnel.[22]

Ten days after the Japanese attack on Pearl Harbor, pedestrians were barred from the East River bridges at night.[23] They would have been vulnerable in the event of a direct attack on the bridge but, more likely, their movements would have been harder to monitor in the dark. The Port Authority increased security on bridges, which included trying to keep citizens from photographing them.[24] Conceivably, a photograph might allow someone to study a span carefully to determine where to plant explosives. Espionage and sabotage, of course, were more likely threats than a direct attack. The author found no evidence of any damage to East River bridges during World War II. As a footnote to this story, it is fascinating to read newspaper coverage of the war in the European theater of operations and count the number of stories that focused on the destruction, saving, or rapid rebuilding of bridges. It was clearly one way of keeping score or measuring progress.

Through the war years and the late 1940s there was no indication that any unusual repair problems prevented Manhattan Bridge from serving its constituents well. Stories in the public domain about the bridge concerned rather routine and predictable matters: traffic bottlenecks at the Brooklyn end of the span, the need for repainting, and the need to invest in rebuilding the approaches to Manhattan Bridge.[25] The operation of this span during the first forty years of its service gave very little forewarning of what was ahead. But, as any stock prospectus reminds readers, past performance is no guarantee of future results!

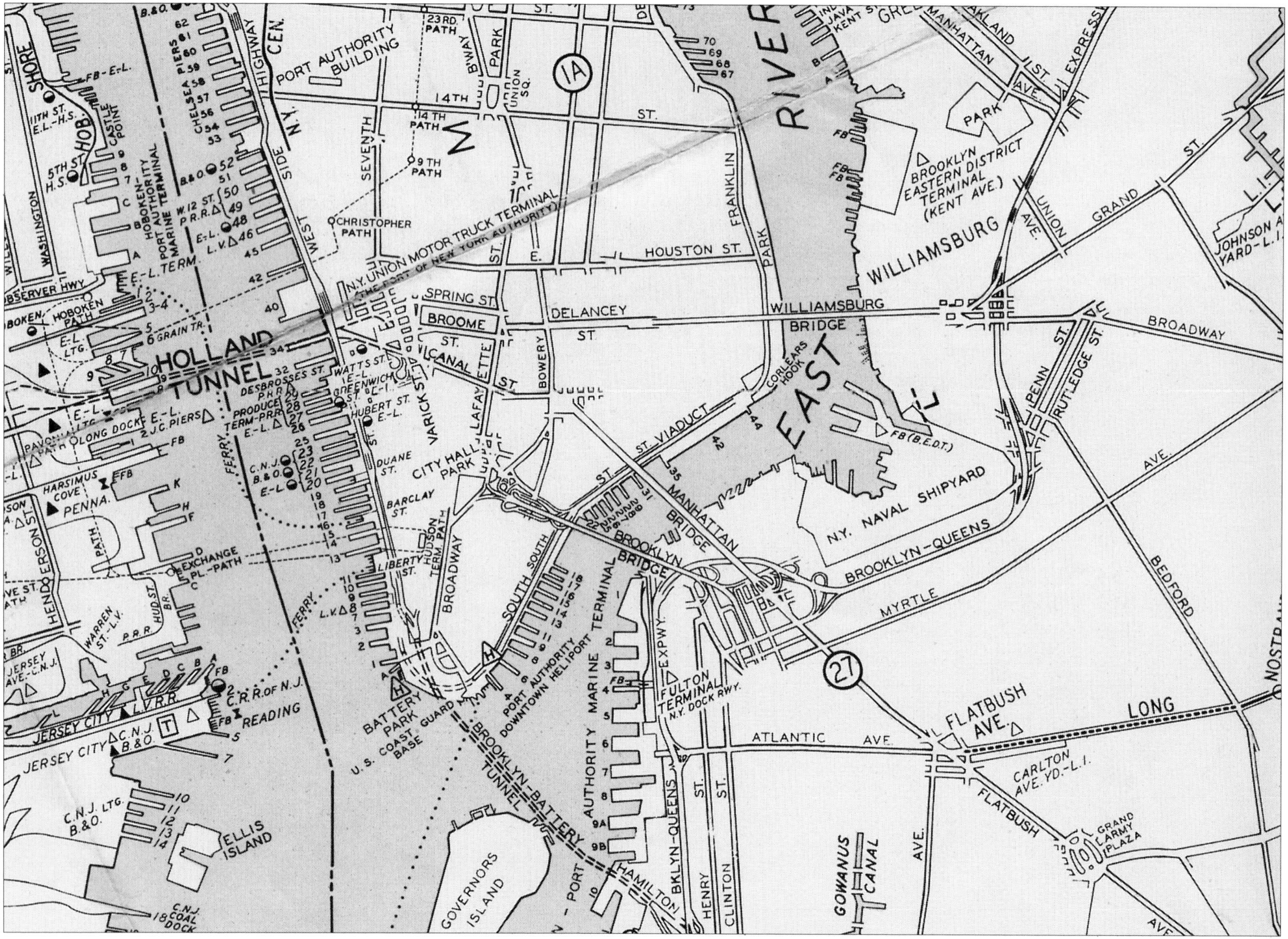

Port Authority of New York and New Jersey map from 1965, showing three East River bridges between Manhattan and Brooklyn.

chapter five

WITHOUT FUNDING, NO MAJOR RECONSTRUCTION

1950–1978

The year 1952 was memorable for many Americans. Republicans rejoiced in the election of Dwight David Eisenhower that restored the GOP to the White House after a twenty-year absence, liberals bemoaned the rampaging investigations of Senator Joseph McCarthy of Wisconsin as he chased alleged Communists, and baseball fans celebrated the 1952 Brooklyn Dodgers, later immortalized by Roger Kahn as *The Boys of Summer*.[1]

That same year Manhattanites and Brooklynites, particularly the commuters, heard foreboding words from Public Works Commissioner Frederick H. Zurmuhlen: "Time and the ruthless demands of the modern age have caught up with the heavily traveled Manhattan Bridge ... so that now the span needs a full-scale investigation of its structural condition.... "[2] Zurmuhlen planned to ask the Board of Estimate for $80,000 to hire engineer David B. Steinman of Steinman, Boynton, Gronquist, and London, who had recently headed a $7,000,000 reconstruction project on the Brooklyn Bridge, to conduct "an extensive study of Manhattan Bridge."[3] The $80,000 request actually had to be approved by Budget Director Abraham D. Beame, a man who later served as mayor from 1974 to 1977.

Zurmuhlen made a telling observation when he complained that the $2,500,000 spent on maintenance on the East River bridges over the past twenty years was far higher than it should have been. In particular, he argued that the Brooklyn Manhattan Transit Company's (BMT's) trains pounding across the Manhattan Bridge placed "a terrific strain on the cables, and all structural parts." He estimated that the average subway train carried 210,000 pounds of human weight, or 105 tons. As each empty car weighed forty tons, a six-car train embodied 240 tons of dead weight plus 105 tons of passengers for a total of 345 tons.[4] The span had not been constructed with this sort of load in mind. It logically followed, therefore, that it might make sense to reduce or eliminate altogether subway and bus traffic on the bridge. Zurmuhlen concluded his argument with the estimate that it would cost about $100,000,000 in 1952 to replace a structure built for $32,000,000 in 1909.[5]

Roughly a month later, Abe Beame and Zurmuhlen were still bemoaning the cost of repairs and reconstruction, with the latter adding that the Manhattan Bridge had consistently generated higher maintenance costs than the Brooklyn, Queensboro and Williamsburg bridges. Zurmuhlen hastened to add that the Manhattan Bridge was perfectly safe, a rote response in a city that had never suffered a major bridge disaster. The most basic problem as defined in May of 1952 was that the BMT was running far more trains on the north side of the bridge than on the south side. These "unsymmetrical loadings" constituted a nightmare for any engineer who studied the span, and yet the BMT obviously scheduled their trains to provide what the BMT believed to be the most efficient commuter service for its customers. Part of the solution, in Zurmuhlen's opinion, appeared to reside in getting the BMT to run more trains on the south side of the bridge.[6]

The Manhattan Bridge was closed to trains temporarily in May of 1953 when a break occurred "in the bottom chord truss on the north side of the bridge near the Brooklyn tower." Commissioner Zurmuhlen hired Bethlehem Steel to do the repair work. At a cost of $33,000, some nine men utilized four tons of steel plates to make the repair in roughly one weekend.[7]

In August of 1953, presumably when most New Yorkers were out of town on vacation and therefore too remote to react and complain, the city announced its allocations for capital improvements to roads, bridges, tunnels, and piers. The biggest piece of the action was proposed by the New York City Transit Authority, which put forth a six-year capital plan that required a figure well in excess of $1,000,000,000, subject to approval by the Planning Commission and Board of Estimate. One of the larger segments of this allocation, roughly $17,000,000 to $20,000,000, was to go for reconstruction of the Manhattan Bridge. The announcement again cited the problem of uneven loading, the study being

Bridge engineer David Barnard Steinman was born in 1886 roughly at the lower Manhattan base of the Brooklyn Bridge. He did his undergraduate work at CCNY and earned a C.E., M.A., and a Ph.D. from Columbia by 1911. After he had taught for four years at the University of Idaho, Gustav Lindenthal invited him to return to New York City and join his staff.

D.B. Steinman. (Smithsonian Institution)

By 1917 Steinman had worked on designing the Hell Gate Bridge and had become a professor of both mechanical and civil engineering at CCNY. Shortly thereafter he joined with Holton D. Robinson in a consulting business. In 1923 Steinman formed the business partnership of Steinman, Boynton, Gronquist and London. One scholar estimates that this firm designed approximately 400 bridges around the globe over the next three decades. Notable projects included the 1931 George Washington Bridge (with Othmar Ammann), the Thousand Islands Bridge between the United States and Canada (1938), and the Mackinac Bridge in Michigan (1957). Between 1948 and 1954 Steinman had the contract for modernizing the Brooklyn Bridge, a job he accomplished while leaving the original external appearance intact.

Steinman enjoyed writing and produced popular books on bridges as well as a few volumes of poetry. The work of his engineering partnership is given some attention is Henry Petroski's *Engineers of Dreams*. One bridge historian claims that Steinman saw himself as the twentieth-century version of John A. Roebling.

conducted by David B. Steinman and Company, and raised a new idea: moving the subway tracks to the center of the bridge.[8]

The suggestion to move the subways to the middle of the span overflows with both sweet reason and common sense. Unfortunately, the wisdom of central placement should have been discerned in the very early 1900s before initial construction was under way. Extensive research into the correspondence between Phoenix Bridge—supplying the structural steel—and Ryan-Parker Construction shed no light on who made this really bad decision. The names Gustav Lindenthal, O.F. Nichols, and Ralph Modjeski are most commonly associated with the question of design, and the approval of the design. The history of the bridge readily reveals several instances when "centering the rails" came to the fore. A great idea, but never acted upon—why? The reason was that moving the tracks to the center actually involved a good deal more than moving the tracks to the center. Any such move had to take into account such considerations as: Where do these tracks lead, and thus how much work is required to alter the plazas and their public art work at each end of the span? What would have to be changed with the approaches on each side? At this point, what sounded like a great idea became horribly complicated and brutally expensive. In a practical sense, in the opinion of the author, it did not turn out to be a viable option.

A TOTALLY NEW PERSPECTIVE ON THE SUBJECT OF MANHATTAN BRIDGE

Most of the discussion to this point has focused on what went wrong with the bridge, and how much were citizens and their political leaders willing to spend to fix it, assuming they knew how. This entire outlook was challenged by John A. Ward's letter to the editor of *The New York Times*, published August 25, 1953. Ward (perhaps a spiritual descendant of sociologist and social critic Lester Ward) contended that major social issues had been overlooked in the discussion over what to do with the Brooklyn and Manhattan bridges. He argued that the development of land in the area surrounding both bridges had been stymied for three quarters of a century. Ward noted, "Bridge spans seriously impede the normal development of land. A walk around the area covered by the Brooklyn and Manhattan Bridge spans should convince almost anyone of this." By contrast, a walk around the areas surrounding the Holland and Lincoln tunnels would create a very different impression.[9]

Ward argued that the city did not realize what it was doing in continuing to support the bridges, for essentially "it is guilty of maintaining one of the most insidious means of creating slums. The majority of the buildings in the vicinity of the two bridges are old and dilapidated." Thus, "under the evil influence of the bridges, these structures will continue to deteriorate." Furthermore, the land that was not developed cost the city a great deal in lost revenue. The answer, according to Ward, was tunnels. They were more modern, did not have a negative influence on surrounding real estate, did not produce slums, and were more secure in the event of bombing raids.[10]

Photo by Berenice Abbott of oyster houses under the bridge at Peck's Slip on South Street, Manhattan, on April 1, 1937. (MCNY)

John Ward's attack on major bridges is somewhat reminiscent of the strenuous criticism leveled against the creation and spread of elevated lines in New York in the late 1870s when an editor of *Scientific American* was moved to observe:

> When the elevated railways ... are completed, four great iron bridges, with numerous branches, will run the length of Manhattan Island.... after people become habituated to trains thundering over them, to thoroughfares blocked with great iron columns, to the impartial distribution of ashes, oil and sparks upon the heads of pedestrians and upon awnings ..., to the increased danger of life from runaway horses and the break-

> ing of vehicles against the iron columns, to the darkening of lower stories and the shading of the streets, so that the same are kept damp long after wet weather has ceased, more such bridges will be erected, and we shall have two storied streets.... The business population ... will be troglodytes—dwellers in dark and shaded caverns.... There will thus be a differentiation ... the results of which students of evolution might ... speculate over.[11]

As if this threat were not great enough, a number of area physicians jointly declared that riding on an elevated line would "induce mental and moral perversion, cerebral exhaustion, insomnia, mania, hysteria, paralysis, meningitis, and decay of nutrition."[12]

Riding on the El, of course, did not generate endless mental or physical problems for commuters, but there were still many good reasons for dismantling most of these iron edifices in the twentieth century. Were great bridges just as bad?

The John A. Ward letter is truly thought-provoking, and surely there must have been more than a few New Yorkers whose humanitarian instincts were aroused by reading it. More importantly, was Ward correct in his reasoning and conclusions? Do great bridges cause blight, slums, and poverty? Or, is it the case that the poor and less fortunate gravitate to the shelter provided by a bridge, and the lower-cost housing that coincidently might be in the same neighborhood? Realistically, there was a lot of substandard housing on the Lower East Side before someone erected a few bridge approaches and anchorages there. Put another way, scholars would agree that the Manhattan side of the Manhattan Bridge could never be mistaken for Park Avenue. What Ward should remember is that two things that coexist in time do not necessarily assume a cause-and-effect relationship. Thus, poverty exists in great cities, but great cities do not necessarily cause poverty—rather, the poor frequently migrate there in search of jobs, shelter, and charity. Similarly, bridges and poor housing may coexist, but that is not the same as demonstrating that bridges cause poor housing. Realistically, the poor have to live somewhere.[13]

Commissioner Zurmuhlen did not buy Ward's "tunnels as a solution" argument, primarily because Ward incorrectly assumed that two tunnels could replace a few major bridges and somehow have the same carrying capacity. Zurmuhlen argued, in a letter to *The New York Times* that appeared September 1, 1953, that soon the Brooklyn, Manhattan, Williamsburg, and Queensboro bridges would provide thirty-one traffic lanes and eight rapid-transit lanes. Twenty tunnels—two lanes each—would have to be built to equal the bridges' carrying capacity, he pointed out. The commissioner placed the cost of each tunnel at $60–70,000,000, which meant that the total cost for twenty tunnels would run somewhere between $1.2 and $1.4 billion. He wondered where the city would get that kind of money.[14] Perhaps Zurmuhlen was overstating his case a bit, but he did feel the need to respond, and he did have a point.

Ironically, less than a year later, as Zurmuhlen listened more and more to consulting engineer David B. Steinman, the commissioner began to see the wisdom of removing the subway trains from the bridge completely and building a separate tunnel for them. Not surprisingly, the issue was money. For $80,000 Steinman had studied the condition of the Manhattan Bridge and put forth two recommendations. The first, as just noted, required a new $90,000,000 tunnel that Mayor Robert Wagner (1954–65) quickly rejected as financially impractical. The less-desirable alternative was to spend $30,000,000 to reconstruct the bridge with the major alteration being the repositioning of the tracks to the center of the bridge. By moving the heaviest loads to the center it stood to reason that much of the torque that was over-stressing the structural steel in the outer lanes would be eliminated. For financial reasons, then, the second recommendation seemed to be a more likely outcome than the first.[15]

The debate over the future of the Manhattan Bridge continued for some time, perhaps proceeding from the assumption that when you remain at the discussion stage, at least you were not spending money. Zurmuhlen and Steinman favored a tunnel while Mayor Wagner and the Board of Estimate at least claimed to favor the $30,000,000 reconstruction (most likely, the mayor and the board preferred a figure under $10,000,000, and their favorite number was zero). On Wednesday evening, January 26, 1955, Commissioner Zurmuhlen took his case to the public over the airwaves on municipal radio station WNYC and condemned the short-sightedness of the Board of Estimate by arguing that if any reconstruction were implemented that kept trains on the bridge, then the rebuilding would last only a few years and would soon have to be repeated.[16] To the commissioner's credit, he never used the public safety issue as a "scare tactic" to get what he wanted. The most he would say was that without action, the span would be imperiled ten years hence.

FINALLY! SOME WORK TO BE ACCOMPLISHED

One job that could move ahead without endless wrangling was a $2,198,169 contract let to John A. Roebling's Sons of Trenton, New Jersey (famous for their work on the Brooklyn Bridge), for replacing cable bands and bridge suspenders, essentially a cure for sagging roadways. In forty-six years of usage "some of the bands attaching the suspenders to the main cables had slipped, resulting in uneven loads on the cables." This repair work would restore the arc to the roadways. Some of the cable bands and suspenders had already been replaced in the late 1930s.[17]

Overseeing this job was Thomas Mombelly, resident engineer of the Department of Public Works, aided by eager civil engineering students (cheap labor) from Brooklyn Polytechnical Institute looking for practical experience as they utilized dynamometers to measure load stresses. Commissioner Zurmuhlen informed the press that this work by Roebling had to be done regardless of how the other debate turned out. He emphasized that the Manhattan Bridge was the second most heavily traveled of the East River bridges, carrying 80,000 vehicles daily. Only the Queensboro Bridge, serving 101,000 cars, trucks, and buses every 24 hours, was more heavily used.[18] The work conducted in March and April of 1955 required a variety of closings. Some roadways were closed from 10:45 p.m. to 5:15 a.m. for a few weeks, other roadways closed for three months. Remembering the closings that were part of the repair schedule was one of the delights of modern metropolitan life.

When the work was completed, the Roebling firm had replaced 584 cable bands and 854 suspenders at a cost of roughly $2,600,000 as opposed to the original figure of $2,198,169.[19] Concurrently, discussions of $90,000,000 tunnels and $30,000,000 reconstruction projects seemed to vaporize, suggesting far more modest expenditures dovetailed with the political and economic realities of the 1950s. This would only change at the point at which the federal government committed to picking up eighty percent of the tab for bridge rebuilding.

150 DARING WORKERS

While Robert Wagner and Abe Beame wrung their hands and dragged their feet rather than acting on the proposals of consulting engineer David B. Steinman and Frederick Zurmuhlen and spending $30,000,000, the basic safety of the Manhattan Bridge was looked after by a team of 150 men who cared for fifty city-owned bridges. Prior to World War II the Brooklyn Bridge alone had a maintenance staff of 200 full-time workers. Officially, these men functioned as inspectors, painters, carpenters, and steelworkers. Unofficially, they performed a staggering number of chores.

Senior inspector Teddy Belziner walked the Brooklyn Bridge, climbed towers and cables, and searched for faults and cracks for roughly fifty-two years between 1904 and 1955. Ed Flood walked the Williamsburg span 300 to 400 times per year for over forty years. He looked for holes in the pavement and inspected bridge members that had been struck by vehicles. Flood examined girder flanges worn down by vibration and by the expansion and contraction generated by temperature changes. He kept an eye out for popped rivets, a common result of a massive subway train jamming on the brakes for a sudden stop. Flood climbed towers with the aid of special shoes and gloves, and tapped rivets with a ball-peen hammer, listening to each rivet's ring to determine whether it was loose or tight.[20]

The bridge inspectors served many other purposes besides testing for all the basic engineering and safety concerns. Ed Flood and his colleagues spent abundant time trying to talk "jumpers" into not jumping. Occasionally one of their own men would "freeze" while climbing. At that point the bridgemen would try to position a man both above and below the stricken bridgeman and lower him to safety. Bridge inspectors walking or climbing a span were frequently viewed by passing motorists as "jumpers" and cars would stop and tie up traffic as well-intentioned motorists tried to talk the inspectors out of jumping. Stray dogs and cats wandering onto a bridge were very dangerous as they led motorists to swerve and cause accidents. Plummeting icicles could destroy windshields and cause accidents as well, so the bridge workers tried to knock these down with wooden clubs before the icicles hit a moving vehicle.[21] The 150 daring men risked their lives climbing a few hundred feet along bridge cables, but also performed remarkably mundane chores as well.

A BIZARRE ACCIDENT DAMAGES MANHATTAN BRIDGE

On June 25, 1958, *The New York Times* reported a collision of two vessels in the East River. Presumably on June 24, the *Empress Bay* (an oil tanker) collided with a freighter, the *Nebraska*. On impact, or shortly thereafter, the tanker exploded, burned, and ultimately sank. A total of thirty-seven crewmen were injured. Alas, the Manhattan Bridge was in harm's way and caught fire. Both the Manhattan and Brooklyn bridges were closed and thus BMT subway service was interrupted. A section of the East River was closed to traffic for a time, as raising and salvaging the *Empress Bay* was no small task. The great conflagration also damaged a fireboat and barge that had been engaged in rescue efforts. A tugboat crew was included among those that had to be rescued.[22] From the vantage point of those charged with caring for the bridge, there were no particularly useful lessons to be learned from this tragic accident that captured the reading public's imagination for several days—unless one wanted to facetiously advise bridge builders to avoid building over exploding oil tankers.

REPAIR AT LONG LAST, OR, $7,000,000 AIN'T $90,000,000 OR EVEN $30,000,000

By October of 1959 New York officials were finally ready to address the long-standing problems of the Manhattan Bridge, at least to the extent that $7,000,000 could purchase solutions. There would be no tunnel or tunnels built to relieve the span of the weight of "345 tons plus" subway trains, nor would the tracks be moved to the center of the structure, but $7,000,000 could buy something. The emphasis of this work was on the politically attractive idea of upgrading service rather than on reconstructing, retrofitting, or repairing. The announced purpose of the spending was the strengthening of two upper roadways so they could serve up to twenty-ton trucks where they had previously served only lighter-weight pleasure vehicles. The project was also going to reconfigure the ornate plazas at each end of the span so that motorists on the Manhattan side could more readily connect with the Lower Manhattan Expressway, and drivers on the Brooklyn side could more readily join up with the Brooklyn–Queens Expressway. A host of lesser jobs would also be completed.[23]

THE POWER BROKER

Robert Moses. (MCNY)

Would anyone dare to demolish or relocate public art in New York City? The basic answer to this question was no, perhaps with one exception. The one obvious exception was Robert Moses, builder of roads, parkways, bridges, parks, beaches, and creator of the New York World's Fair, labeled simply in Robert Caro's classic 1,162-page biography: *The Power Broker.* Without engaging in overstatement, Caro concluded that the Yale- and Oxford-educated Moses " ... held power, a power so substantial that in the fields in which he chose to exercise it, it was not seriously challenged by any Governor of New York State or by any Mayor of New York City."[25] In other words, " ... Robert Moses shaped New York. Physically, any map of the city proves it."[25]

The power broker, in a rather naked display of influence in June of 1961, asked the New York City Art Commission for the right to demolish both ornamental architecture and statuary that were part of the Brooklyn and Manhattan plazas of the Manhattan Bridge in order to make way for rebuilding.[26]

Whether Moses actually intended to destroy the public art, or was just making a tongue-in-cheek pronouncement in order to enjoy the reaction is not clear. But, of course, the reaction came. The Brooklyn Museum of Art immediately took a public position in opposition to demolition, and a few days later *The New York Times* editorial page concurred.[27] Ultimately a compromise was worked out whereby the public art was to be relocated to the Brooklyn Museum and the Bronx Park Zoologic Gardens.[28] The transfers were completed by the mid-1960s.

Canal Street and the plaza at the Manhattan end of the bridge, May 9, 1917.
(Collection of The New-York Historical Society)

Above: Statues by Daniel Chester French at the Brooklyn end of the bridge represent Manhattan and Brooklyn. November 1, 1916. (Collection of The New-York Historical Society)

During the tumultuous 1960s, the uproar generally did not extend to the Manhattan Bridge. Protests of all sorts that predictably sought maximum publicity exploited the world-famous Brooklyn Bridge only a few blocks away. One rather humorous episode in April of 1964, however, involved a rock that fell from the Manhattan Bridge during renovation and struck and injured a passenger in a boat passing under the span. The owner of the vessel complained to the New York City Public Works Department. The department representative, recognizing that the odds of such an occurance were probably 10,000,000 to 1, defended the department by pointing out that such accidents did not occur with great frequency.[29] For the remainder of the decade public attention was rarely drawn to the span apart from an occasional newspaper story—on a slow news day—pointing to the ever-present problem of traffic congestion.

MORE TROUBLE AND PUBLIC DEBATES IN THE 1970S

On November 24, 1970, the commuting public in New York learned that at least two lanes of the Manhattan Bridge were going to be closed for four months in order to complete emergency repairs. The saltwater from the estuary below, time, rain and snow, dirt, and pigeon droppings had produced severely deteriorated expansion joints in the bridge. In theory, these joints were to have been cleaned regularly. In practice, they became full of crud and rust.[30] (Understandably, some modern spans are designed to function without expansion joints.)

The Clean Air Act passed by the United States Congress in 1970 provided the Environmental Protection Agency with the impetus to identify new means of improving air quality, admittedly in a cultural setting that appeared to be increasingly sensitive to pollution of all sorts. To that end, a 1975 EPA directive called for required tolls on the East River and Harlem River bridges by mid-1977 that, in turn, would reduce the number of vehicles using the bridge—thus lowering levels of air pollution. The Beame administration moved quickly to challenge this EPA directive in federal court, and noted the distinct possibility that tolls would not necessarily reduce traffic significantly. During 1975 and 1976 EPA officials did their best to pressure the appropriate officials to institute tolls. What the federal agency failed to understand was that toll-free bridges, similar to the once-universal nickel transit fare, were seen as a sacred entitlement by many New Yorkers who frequented East River bridges. Thus, while cleaning up the environment no doubt could be viewed as an idea whose time had come in the 1970s, the local political wisdom militated against tolls.[31]

What the EPA could not accomplish, the local Brotherhood of Teamsters could and did: namely, reduce pollution on the bridge. The "bridge tenders" struck for higher wages in January of 1972, and the East River bridges were shut down.[32] These labor troubles flared intermittently through April of 1973.

On August 19, 1977, New York City and the federal government announced that they had reached a clean-air accord for Manhattan that would "rely less on banning parking and more on bolstering mass transit and enforcing traffic laws." In place of the traffic ban, New York was "promising to eliminate illegal parking south of 59th Street, establish priority bus lanes ..., set up "park and ride" programs ..., promote staggered work hours, ... and take all actions feasible to keep the 50 cent fare." Mayor Beame contended that this compromise would protect both the air quality and economy of Manhattan. He further noted, " ... the proposal to put tolls on the East River bridges to reduce the number of cars entering Manhattan has been dropped...."[33] Alas, the sanctity of the toll-free East River bridges had been preserved!

The Brooklyn Bridge and one World Trade Center tower are on the far left in this 1978 aerial view looking toward Manhattan. (Photograph by Jack Boucher for Historic American Engineering Record)

chapter six

TAPPING THE FAUCETS IN WASHINGTON, D.C.

Federal Dollars Aid Manhattan Bridge, 1978 to 2000

The debate that the Abe Beame (1974–1977) and Ed Koch (1978–1989) administrations had with the EPA (particularly during the Carter era) over the nexus between clean air and bridge tolls was a timely reminder of the potential for the growing importance of federal influence in the life of New York City. Consider the contrast between the newspaper story of March 27, 1978, that reported the city "has replaced 250 of 300 weakened steel floor beams on the Manhattan Bridge"[1] and Mayor Koch's announcement the following March 20, 1979, that $244,000,000 would be spent to rehabilitate 126 city bridges.[2] The quantum leap so evident here is explained by the infusion (or should it be "deluge"?) of federal dollars such that $195,000,000 or eighty percent of the cost of rehabilitation was being paid for by Washington under the Federal Bridge Replacement Act of 1978 and was targeted for the rehabilitation of several noteworthy New York spans. Concurrent with this source of funding was the beginning of an extended relationship between Manhattan Bridge (with all its aches, pains, and design problems) and Weidlinger Associates civil engineering firm of New York City. Weidlinger not only guided the reconstruction of the span between the late 1970s and late 1990s, but currently (2003) holds major contracts to continue the work for at least another decade.[3] From time to time the engineering department at Columbia University has displayed interest in the design quirks of Manhattan Bridge, and has even gone so far as to construct a model for teaching and testing purposes—a good negative example.

MAYOR KOCH'S "MODEST PROPOSAL" OF 1980

Perhaps as an Independence Day gift to the citizens of New York, on July 3, 1980, "hizzoner" announced an "experiment" or "three month test" that would ban all single-occupancy cars from East River bridges during morning rush hour. At first consideration, the plan seemed to make perfectly good sense, delighting at least those New Yorkers who had no need to commute. The mayor was applauded widely by environmentalists and bicyclists, the latter being generally accustomed to the concept of single occupancy. What Ed Koch appeared to have overlooked, however, was a host of other interests.

For example, the American Automobile Association filed suit as this ban obviously infringed on the rights of motorists. The Metropolitan Garage Board of Trade, which saw dollars flying out the window of its members' garages, also registered its opposition. These two organizations gained fervent support from the National Council of Salesmen —representing people who frequently conducted business as single occupants of a vehicle. Mayor Koch was reminded very quickly and forcefully that there was more to running the Big Apple than pleasing environmentalists and a few other well-meaning but disinterested parties.[4]

Let the record show that three weeks later the New York Superior Court blocked the mayor's ban.[5]

The Manhattan approach to the bridge is seen here, with heavy traffic in both directions. Midtown is in the background. (Photograph by Jack Boucher for Historic American Engineering Record, 1978)

FEEDING AT THE FEDERAL TROUGH GETS BETTER

By the time that President Ronald Reagan took office in early 1981 there had been enough studies conducted by the United States Department of Transportation and others, and enough data amassed, to convince almost anyone who was interested that the infrastructure of the United States in general, bridges in particular, was in need of serious attention if disasters were to be averted. Bridge historians also sounded the alarm as did the Society for Industrial Archeology, the Society for the History of Technology, and the American Society of Civil Engineers. Any concerned citizen knew there was a problem. One of the better recent statements has been produced by Clark Wieman in the January 1993 issue of *Technology Review*, the M.I.T. alumni magazine. In an article entitled "Road Work Ahead: How to Solve the Infrastructure Crisis," Wieman estimated that "about 40 percent of U.S. bridges are rated deficient. An average of 120 collapse each year."[6]

Perhaps it logically followed that New York City officials on June 30, 1981, proclaimed a commitment of $115,000,000 over ten years to reconstruct four lower East River bridges.[7] In a surprising development roughly three months later, Secretary of Transportation Drew Lewis and the Reagan administration made a particular pledge to pay the entire cost of rehabilitating both the Manhattan and Williamsburg bridges. This decision was made in opposition to the extraordinary argument of an attorney for the Federal Highway Administration, who wanted to withhold monies that would go to reconstruct the subway-serving portion of the bridges.[8]

On May 25, 1982, bridge inspectors found cracks in a twelve-foot steel beam that was part of Manhattan Bridge. The city that had never suffered a major bridge disaster obviously did not want to tempt fate, and thus the span was closed to subway traffic to and from Brooklyn for several days while repairs were made. Commuters altered their subway routes and got to work with one or two additional transfers. This dislocation does not sound like much of a problem until you see photographs in newspapers of passengers crushed together like cattle trying to make their way through the mob, pursuing new paths with a modest degree of uncertainty—not quite sure where they were going, not quite sure they could get through the crowd in time to try to board an already jam-packed train.[9]

RONALD REAGAN, A NEW GAS TAX, A FILIBUSTER, AND RENEWED SUPPORT FOR HIGHWAYS AND BRIDGES

On the eve of Thanksgiving Day 1982, when the mind of the nation was not focused on Washington, D.C., President Reagan proposed a new five-year plan to spend $32.9 billion repairing the nation's bridges and highways. As in the past in most instances, the federal goverment stood ready to pay eighty percent of the cost of these projects. Controversy arose, quite predictably, when the president also proposed a five-cent-a-gallon increase in the federal tax on gasoline to generate the $32.9 billion. The highway lobby, led by the nation's auto makers and oil industry, to name a few, was not amused. Not surprisingly, the legislation was met with a filibuster in the United States Senate, but ultimately was approved.[10]

Few, with the exception of pioneering modern photographer Berenice Abbott, treated the Manhattan Bridge as a great cultural icon—the sort of iron and steel monument that inspired endless poets, writers, song writers, and artists. That grand piece of architecture was just a few city blocks downstream. It was refreshing, therefore, to find even a modest example of New Yorkers engaged in a simple appreciation of Manhattan Bridge. On June 26, 1983, at 1 p.m. the 92nd Street Y sponsored a tour of Lower Manhattan that featured a visit to the span and a talk by Michael George, who explained the planning and construction of Manhattan Bridge together with the changes it had undergone during its first 74 years of serving the public. The tour cost $6.50 and was scheduled to proceed rain or shine![11]

Returning to reality, *The New York Times* at the end of the summer of '83 announced: "Subway riders, already suffering from a summer of bad service, will encounter a new delay Monday when trains that run on the Manhattan Bridge are rerouted to make way for the construction work. The 74 year old bridge over the East River has deteriorated dramatically over the last few years.... Work on the bridge's four subway tracks will continue for about three months. During the first six weeks, the BMT's N and QB trains will be diverted. The trains will instead be sent through heavily used East River subway tunnels."[12]

THE WEIDLINGER INFLUENCE

Any reasonable person, looking at the maintenance record of Manhattan Bridge through the last half-century, might be tempted to compare it to one man's definition of history: one damn thing after another. For years, cracks appeared, steel deteriorated, repair monies were sought and obtained, and a contractor hired to make the repairs. Almost everything was done on an ad-hoc basis.

Beginning in 1978, however, the rickety road connecting Canal Street to Flatbush Avenue entered a new and modern phase for essentially two reasons. First, federal legislation that year, and ongoing support during the Reagan years and beyond, meant that Washington would be willing to shoulder between eighty and one hundred percent of the cost of all major reconstruction efforts. Second, with funding in place, a long-term relationship was launched that linked the bridge to Weidlinger Associates, Inc., a highly regarded civil engineering consulting firm with home offices in New York City. In practical terms what this relationship meant was that the full weight of modern engineering would be brought to bear on analyzing the structure, and then a long-range plan for reconstruction would be mapped and carried out. The days of patch, patch, patch had finally come to an end.

From Weidlinger's perspective:

> Our engineers first inspected the bridge in 1978 and found persistent cracking and corrosion.... [A] successful cure depended on establishing a cause of the cracking. Computer analysis showed the cause, and why previously proposed methods ... couldn't stop it.... [W]e developed a repair scheme that promised to correct the structure's design deficiencies, restore its lost capacity, and extend its life span at least fifty years. Our report ... presented a plan that would take eight to ten years to complete while maintaining most of the bridge traffic. The estimate was ... conservative. More than fifteen years later (1996), the Manhattan Bridge rehabilitation is in its final phase.[13]

DEFLECTION, CRACKING, FATIGUE AND TORQUE TUBES

Weidlinger engineers quickly understood that lengthy and terribly heavy subway trains traveling along this two-level, seven-lane bridge on the outside lanes essentially twisted the span. A train or multiple trains crossing the bridge on one side created a deflection of anywhere from two to eight feet more than on the opposite side. This torque along the longitudinal axis of the span occurred with great frequency during peak commuter hours when trains may have been crossing every four minutes. The twisting or torque created cracking. Cracks began as something almost invisible to the human eye and, as a result of metal fatigue, continued to grow until they became dangerous. (Try a small-scale experiment yourself: If you twist one end of a paper clip three or four times, metal fatigue will cause a piece to break off.)

Weidlinger, led by Herbert Rothman, brought a dual solution to this torque and cracking problem circa 1982 by first reinstalling far stronger lateral braces than the span had ever had before, and by creating what they called a "torque tube." This was achieved by placing "massive diagonals under the upper roadway" that, in turn, helped the Manhattan Bridge to resist twisting. The consulting engineers claimed that the torque tube may have diminished twisting by as much as fifty percent.[14]

The Weidlinger team believed that they were actually looking at more deterioration than the deflection-inspired torque by itself might have explained. The other source of difficulty they identified was something they chose to call "*stick–slip* behavior of partially frozen expansion bearings between the roadway framing and the suspended structure. The *stick* multiplied the twist-induced fatigue eightfold, while the *slip* ground the bearings down."[15]

AN INNOVATIVE 108-FOOT TEST SECTION

Harboring the conviction that the cracking issue was at least under control, in 1982 the consulting engineers next turned their attention to creating a 108-foot test segment of the span. This new idea allowed them to test just how good their analysis and repair plans were. The practical problems associated with installing this panel provided Weidlinger with some sense of the difficulties associated with working on the Manhattan Bridge and thus better enabled them to assess costs. Other engineers were sufficiently impressed with the concept of a test panel to apply the practice to the rehabilitation of other large bridges, including the Golden Gate Bridge.

The test section was also fitted with elastomeric bearings to replace the expansion bearings that had long since frozen, and ruined lower roadway was replaced with modular panels. The strain on the test segment of the bridge was measured while two substantial subway trains were on the span. When the engineers determined that both the calculated and measured stresses concurred, they were reasonably confident that their repair plan was workable.[16]

ADDITIONAL REPAIR WORK IN THE 1980s

Weidlinger Associates contend that emergency work on the three other lower East River bridges in the mid-1980s precluded a good deal of work that might have been carried out on the Manhattan Bridge. Still, some matters had to be attended to such as repairing "floor beams, stringers, and trusses." In addition, disc bearings replaced pin-and-link connections that held the suspended structure and towers together. Furthermore, in the space of a year, "three quarters of the wires in one main cable were cut, socketed, and reconnected to new girders which were tied deep into the existing concrete anchorage with steel rods."[17]

The consulting engineers were apparently a bit shocked or stunned when they lost their contractor, the American Bridge Company of Ambridge, Pennsylvania, before they had fully completed one of three contracts they held for the bridge. The contractor claimed to be redefining its business. The initial agreement had been worth $40,000,000 and provided for reconstructing or stiffening the side spans. The new contractor was stymied in proceeding when it was learned that much of the steel stockpiled for the job was rusted, stolen, or perhaps both rusted and stolen. When the first contract was completed, work under a second agreement was initiated sometime in 1992 and carried a price tag of almost $100,000,000 to complete the "torque tube steel work" on the west portion of the span.[18]

Between 1995 and 2005 a third phase costing $350,000,000 is envisioned. This does not mean, however, that anyone can actually foresee an end to maintenance and reconstruction on Manhattan Bridge. Phase three work includes: "... stiffening of the suspended main span ..., steel framing on approach roadways ... and new expansion joints and bearings and new overlays for roadway surfaces." In addition, the lower roadway will be replaced by a "concrete-filled grid deck system."[19]

DOES A GOTHAM SPAN NEED TO BE PROTECTED AGAINST EARTHQUAKES?

Herbert Rothman and Weidlinger Associates continued to pour their time and energy into the Manhattan Bridge, but that is not to say new concerns cannot surface. In Weidlinger's 1996 newsletter the job of "retrofitting" the span to survive in the face of an earthquake was raised for the first time. Surely it is possible to address a "seismic upgrade," but what degree of security should be added? At what cost? Is the geology of metropolitan New York the same as the geology of the San Francisco Bay area? (In the early 1990s roughly $128 million was spent on the Golden Gate Bridge to add "stiffening structures" and giant "shock absorbers" in order to address the "seismic energy" in the Bay area.)[20] Obviously the two settings are quite different as New York City lacks proximity to a major fault line. What degree of earthquake should the bridge be protected against? What portion of the span should be expected to survive such a disaster? How effectively does anyone retrofit a ninety-year-old structure? These are difficult questions and the consulting engineers were not optimistic. They concluded their 1996 story by noting: "Unfortunately, the repair cost grows with each passing year, as unimpeded and relentless fatigue and corrosion continue to take place in the still unrepaired structure. Despite extreme efforts on the part of everyone involved, work will not be completed until well past the year 2000." [21] Indeed, it is tempting to ask, rhetorically, whether the concept of "work completed" and "Manhattan Bridge" should be used in the same sentence.

THE NEWSPAPER OF RECORD TAKES A PRONOUNCED INTEREST

During the 1990s *The New York Times* has focused considerable attention on the need for bridge reconstruction. Perhaps because the editors sensed that, where the East River bridges were concerned, the city was living in the "eleventh hour." Or perhaps because of the shock value of the failure of a New York State Thruway bridge in the late 1980s, when the Schoharie Creek reached its historic high. In any event, the *Times* coverage in the 1990s has been commendable.

A good example of this was a subtle yet devastating piece written by Calvin Sims and printed on December 17, 1990. A modest title on the bottom half of page one, column one, read "Corrosion's Rise Imperils Bridges Over East River." Deputy Transportation Commissioner for the city's Bureau of Bridges, David Steinberger, warned, "If this trend continues, dozens of bridges will be closed next year." But, this was rebutted by Transportation Commissioner Lucius J. Riccio who reassured all that the situation was manageable.[22] However, buried in Section B, page five, column one, was the rest of the story, featuring a chart that identified the number of red and yellow "flags" awarded to East River bridges between July 1 and December 14, 1991. A red flag denotes a serious problem that could lead to a partial or total bridge collapse. A yellow flag denotes a problem that could soon warrant a red flag.

EAST RIVER BRIDGES AT RISK

	Red Flags	Yellow Flags	Total
Manhattan Bridge	288	177	465
Williamsburg Bridge	48	212	260
Queensboro Bridge	12	12	24
Brooklyn Bridge	7	9	16
All other NYC bridges	61	182	243

Source: New York City Department of Transportation

The data on flags are alarming, particularly for the Manhattan Bridge, but become even more so with the recognition that a red flag a decade earlier would have triggered a response in a few hours or a few days at the most. In 1990 the response time, realistically, was closer to six weeks.[23] Thus, in a mere ten years, the city that had never suffered a major bridge disaster markedly reduced its margin of error.

Two other observations from the same story were equally chilling. First, referring to several thousand large cracks in the support beams of the Manhattan Bridge, the author noted: "Many of the cracks were discovered two years ago, and at that time engineers drilled holes in the paths of the cracks to stop them.... [I]nspectors have found that the cracks now extend from the other side of the holes."[24] Second, three sketches of the end view of an "I" beam illustrated the impact of corrosion from 1903 to 1984 to 1988, with the thickness of the web flange tapering from three-eighths to one-quarter to zero inches, demonstrating the accelerating speed with which corrosion destroys a steel beam.[25]

THE PARADOX OF RESTORING SUBWAY TRAINS TO THE SOUTH SIDE OF THE SPAN

In a less-than-perfect world the basic paradox was this: the more authorities gave in to public pressure to restore the N line, with its 25,000 constituents, to bridge use (for example), the more the added use would guarantee additional bridge shutdowns. However, since the south side had been shut down frequently, officials found it harder and harder to argue that current inconvenience would lead to a better tomorrow. Accordingly, in September of 1990, Transportation Commissioner Lucius Riccio restored N train service to the south side of the bridge, despite opposition from Deputy Commissioner for Bridges David Steinberger, who pointed out that more red flags would soon be required.

Commissioner Riccio was fully aware of the paradox just cited, but he also had good arguments of his own. He wanted better service to Brooklyn, and he wanted to encourage people to abandon their cars in favor of mass transportation. What Riccio understood was that if riders turned their backs on subway travel for too long a period, they might never return.[26]

IF MORE IRON AND STEEL FAIL TO RESTORE THE BRIDGE, TRY WOOD

The need for more and more emergency closings of the span only seemed to grow through the 1990s. On Friday February 15, 1991, Calvin Sims reported: "For the fourth time in less than two months, a major section of the badly deteriorated Manhattan Bridge will be closed for emergency repairs ... disrupting weekend subway service on the B and D lines ... to repair large cracks in steel support beams underneath the tracks and roadway on the bridge's north side."[27] Yet a new problem surfaced in this same story as train operators on the B and D lines reported loose pieces of wooden scaffolding dangling perilously close to the trains. The cure for this was additional wooden scaffolding to brace the wood already installed.[28]

The solution seems to make perfectly good sense, but it does help to remember that wood is rarely bolted or riveted into place. Furthermore, one of the great truths of modern urban life is that anything not bolted down has a way of moving—sometimes great distances. This observation leads into another Calvin Sims story for February 27, 1991, that summarized the results of an investigation authorized by New York City Council. The investigators dutifully collected some 11,000 pages of documents and essentially determined that there were now gaping holes in the decks and steel beams of the city's bridges, and that an abundance of wooden pillars had been employed to support the rusting structures. The newest problem was that, allegedly, the homeless were stealing the beams and using them for firewood to cook and keep warm, thereby imperiling already weakened structures.[29] Did this theft constitute a criminal act?

In 1991 New York City was still basking in its modern ethos of "cosmopolitan and sophisticated people are, above all else, tolerant." Accordingly, under Mayor David N. Dinkins, squeegie men and graffiti artists were just "creative" citizens contributing to the metropolitan ambiance, and homeless people stealing support beams from bridges should certainly not prompt value judgments. Thus, dismantling makeshift bridge supports was more likely to be viewed as a human right than destruction of public property.

SPOTLIGHT ON CORROSION AND DECAY

On February 28, 1991, a highly publicized hearing was scheduled to be held in City Hall for the purpose of analyzing the findings of the investigators. Basically, the question was whether the city's bridges could survive another cut in spending on bridge maintenance by the Dinkins administration. The hearing would feature videotape that highlighted the extensive corrosion that had been uncovered, and one could expect this videotape to find its way into the hands of area television stations.

Equally as dramatic would be the conflicting testimony offered by Commissioner Riccio and his *former* deputy David Steinberger. The latter had been fired in January of 1991 for publicly criticizing his boss. Steinberger had openly challenged the city's budget for bridge maintenance, deeming it inadequate, and had sided with transportation department engineers who sharply opposed Riccio's decision the previous September to restore subway service to the south side of Manhattan Bridge. To add to the plot, the proceedings would also test the power of City Council versus the Dinkins administration under the new City Charter. *The New York Times* made its position clear by including a subheadline in the story: "Ammunition for a hearing on mismanagement."[30]

NEW YORK STATE INSPECTS MANHATTAN BRIDGE IN AUGUST OF 1991

In response to all of the public attention drawn to the span, in the summer of 1991 the state hired Steinman, Boynton, Gronquist & Birdsall, a Manhattan-centered engineering firm, to conduct its own investigation. The results were even more alarming than those cited in the previous chart. Thus, where the department of transportation data for 1990 identified 288 red flags and 177 yellow flags, Steinman and Boynton awarded 379 red flags and an even greater number of yellow.

Most fascinating was the discovery of a hole the size of a quarter in a rusty beam—a discovery that shut down subway service over the bridge to 250,000 rush-hour commuters on Wednesday evening, July 29, 1991. To put this development in a positive light, Martin Gottlieb of *The New York Times* observed: " ... engineers and officials from three government agencies were almost unanimous ... in their view that the inspection system which costs millions of dollars and involved two sets of consultants and a quartet of city inspectors, proved its worth when the hole was detected."[31]

October 1991 view of the Manhattan tower and anchorage with the twin towers of the World Trade Center in the left background. (Photograph by Jet Lowe for Historic American Engineering Record)

WOODY HARRELSON AND "THE COWBOY WAY" ENLIVEN THE BRIDGE

For a host of reasons, major cities enjoy providing the backdrop for major motion pictures. In 1993 New York City hosted the filming of "Carlito's Way" and a bit later "The Cowboy Way," the latter being a story of two rodeo cowboys from New Mexico who come to New York City to search for a missing friend. Both productions required use of the subway system, and the filming of "Carlito's Way" unfortunately generated serious tension and ill will between the City Transit Authority and the Hollywood visitors. Thus, shooting "The Cowboy Way" provided an opportunity to mend fences with the industry.

In an effort to create some drama never seen before on film, the script called for Woody Harrelson and Kiefer Sutherland, on horseback, to chase a subway train onto the Manhattan Bridge in pursuit of a villain. This would be visually interesting, two hundred feet above the water, and not a particularly difficult stunt to stage. The major adaptation required was a plywood platform for two horses to run on, which carpenters constructed in ten days. When time for filming arrived on Saturday November 13, 1993, the subway-car operator was instructed to accelerate to twenty miles per hour, Woody Harrelson and Kiefer Sutherland chased the subway on horseback onto the span, then jumped aboard the train and captured the bad guy. The creative scene was a success. Fortunately, a chronicle of this stunt was reported by James Lardner for *The New York Times* on June 5, 1994, and the reading public was finally able to read something positive about life on the Manhattan Bridge.[32]

CLOSED AGAIN FOR REPAIRS IN MAY OF 1995; ALPHABET SOUP FOR SUBWAY RIDERS

On May 2, 1995, the newspaper of record offered its readers a story of the Manhattan Bridge shutdown to subway traffic on weekdays between 9:45 a.m. and 3:30 p.m. for the next seven months. The ban also applied from 4 a.m. Saturday through 10 p.m. Sunday. The story included a map and seemingly endless additional instructions for the straphangers.

Two recent slides by Thomas Flagg show Manhattan Bridge shrouded in scaffolding. (Thomas Flagg)

Consider the existential pleasure of grasping and acting on the following information: "The B, D, and Q lines will not cross the bridge while work is underway. On weekend days and midday on weekends, the B will only run south of Atlantic Avenue, and the D ... only north of 34th Street. The Q will run on the N and R tracks from 34th Street to DeKalb...." Furthermore, "The M train will go no farther south than Chambers Street ... skipping Brooklyn entirely, and the Z train will not run at all."[33]

Citizens who missed the newspaper stories about the elimination of service or who neglected to cut out the convenient map provided were still able to receive abundant instructions in the subway stations from strategically placed transit workers equipped with bullhorns. Of course, those who actually knew where they were going also got to enjoy the bullhorns.

Manhattan Bridge in the summer of 1997, looking toward the DUMBO section of Brooklyn. The acronym means Down Under the Manhattan Bridge Overpass. It is an area with some light industry and many artists' lofts. (Thomas Flagg)

AND THE BEAT GOES ON

As the millennium approached, the cadence of the narrative in some respects seemed to become more and more predictable. The deeper engineers dug into the aging span, the more problems they uncovered. Layers of corrosion were not really protecting good steel, as one might suppose or hope, but only more layers of corrosion. Accordingly, estimates of time required to complete reconstruction and total costs kept spiraling in one direction: upward.

Gary Pierre-Pierre, in a feature story in *The New York Times* on April 10, 1996, observed that from a starting point of 1982, the completed work would require roughly a decade longer and $300,000,000 more than experts had predicted. And thus from the vantage point of 1996 the predicted end date became 2003 and the total cost became $452,000,000.[34] If this price seems staggering, a year and a half later, in September of 1997, the MTA estimated the total cost as $470,000,000.[35]

As noted earlier, the subway-riding public of New York City and the Manhattan Bridge are very much like two dependent people in a bad marriage—they do not seem to do very well together, and yet they can hardly stand to be apart! On a more positive note, photographer Berenice Abbott, commenting on *Manhattan Bridge Looking Up*, observed: "Every rivet, bolt, nut, screw, and beam girder of Manhattan Bridge, as one looks up at it, comes to life as the portrait of integrated techniques by which civilization survives."[36]

The Brooklyn, Manhattan and Williamsburg bridges over the East River, seen from the top of the World Trade Tower in 1982. (Photographed by Jet Lowe for Historic American Engineering Record)

chapter seven

THE MANHATTAN BRIDGE STORY

Summarizing Thoughts to an Ongoing Saga

In thinking about the broader historical significance of a series of bridges across the East River it is important to remember that in 1900 the Lower East Side was considered by some to be the most densely populated urban setting on Earth, comparable even to Calcutta. Accordingly, endless social problems and human misery abounded, documented skillfully by observers such as Jacob Riis.

One solution resided in a better urban transportation network that would allow people to leave the crowded confines of lower Manhattan but then commute back to the city for work and perhaps even leisure activities. With the coming of a new century there was a shrill public outcry to extend existing transportation arteries, and apparently political leaders heard the outcries. (We later learned that politicians love to cut ribbons on new bridges and parade over them in bombastic opening ceremonies, taking as much credit as humanly possible.)

The Williamsburg Bridge was completed by 1902 and the Queensboro and Manhattan bridges by roughly 1909. Furthermore, the massive New York subway system was expanded to travel over some of these spans. With bridges and subways superseding the overcrowded, less reliable, and sometimes dangerous ferries, literally tens of thousands of Jewish and Italian immigrants (and others) saw this as an opportunity to make their way out of the congested Lower East Side tenements into the more spacious and attractive accommodations of Brooklyn and Queens. The new bridges redefined the configuration and distribution of Gotham's population.

If the Brooklyn Bridge was the crown jewel in Gotham's universe, then the Manhattan Bridge was something of a tarnished bluish rhinestone hovering in the shadows a few blocks upstream. If "The Rickety Road to Flatbush" was not one more great American success story, it still offered a number of significant insights into the history of bridge building and posed daunting questions for those entrusted with maintaining, restoring, and operating aging spans with public tax dollars.

This volume has provided the chronicle of a span whose history almost no one ever expected to see. The Manhattan Bridge in the eyes of some, perhaps, never merited a history. Even though it has always been physically imposing, even though it has been a recognized metropolitan monument for the past ninety years, and even though it serves a few hundred thousand commuters daily, it is rarely—if ever—referred to as a great bridge. Great bridges, it seems, are those created by great designers. Great bridges are those commonly celebrated for their beauty and engineering merit. All the rest, somehow, are far less worthy of attention. Indeed, Berenice Abbott's interest in Manhattan Bridge can easily be interpreted as a statement by a rebel and maverick in the art world who chose to go out of her way to love the unlovable, to care about a civic monument that merited almost no one else's attention.

Joining her quite recently are two others: photographers. Horst Hamman in *New York Vertical* and Stanley Greenberg in *Invisible New York* have both published photos depicting portions of the Manhattan Bridge.[1]

The tale began with the complex and contorted story of construction. A burgeoning metropolis at the turn of the century that needed to upgrade its transportation network found that it took a decade to plan, design, and erect Manhattan Bridge. The distinguished Austrian-born bridge designer Gustav Lindenthal, who had been "paid off" or "bought out" as designer of a suspension bridge at Québec in the early 1900s, was ousted from the Manhattan Bridge project in 1904 when "reform" Mayor Seth Low was defeated at the polls in 1903. Prior to his dismissal, however, Lindenthal had engaged a number of civil engineers in a multi-year debate over the relative merits of eye-bars versus cables for the span over the East River. Lindenthal was hardly a neutral observer, as he had been delighted with the use of eye-bars on an earlier suspension bridge he had designed in Pittsburgh, and was personally at odds with the Roeblings—America's premier producers of wire cable—stemming from a dispute over the construction of the Williamsburg Bridge.

When Mayor George B. McClellan, Jr., entered City Hall he appointed O.F. Nichols chief engineer to replace Lindenthal. This may well have constituted progress Tammany-Hall style, but as far as planning and design were concerned it roughly meant starting over as only Lindenthal's design for the towers was retained. Carrere and Hastings designed the St Denis-inspired portals.[2] As late as 1905, therefore, New York observers were still wondering when the necessary property would be condemned, when homes would be razed, and when the construction of anchorages would begin. Practically speaking, this meant delay and more delay; and, alas, no structural steel pointed skyward until 1906. The public was justifiably outraged, and considered the project scandalous—not because of traditional payoffs, kickbacks, or bribery, but because a much-needed public transportation link had been dragged doggedly through the muck and mire of urban politics.

What the citizenry did not know about this new span was that it had been designed to make some use of Latvian-born bridge designer Leon Moisseiff's "deflection theory," an idea really in its experimental stage in the early 1900s and thus not really perfected until the 1920s. Based on mathematical modeling, deflection theory would encourage construction with less steel and more flexibility. In practical terms, in retrospect, the structure appears to have been underbuilt, particularly for the loads it would be asked to support throughout its lifetime. In some respects, then, the bridge's greatest limitation was unknown to both the engineers and the general public when it was opened on New Year's eve prior to 1910.

The "cowboys of the skies" or structural steel workers who erected the Phoenix-fabricated steel between 1906 and 1909 were obviously workers at great risk. Tossing and catching white-hot rivets a few hundred feet above the East River was hardly a prescription for good health and longevity. At the same time, the advent of the International Association of Bridge and Structural Iron Workers meant that someone was now making careful note of injury and death on the job, and was working to make this information public—at least through the monthly publication of *The Bridgemen's Magazine*.[3] The magazine reported that many bridgemen made over $5.00 a day at a time when many industrial workers made only half that. The union claimed that seven percent of their workers were injured every week, and provided advice to widows who might have to negotiate with the Phoenix Bridge Company upon the death of a spouse.

Enigmatically, the Manhattan Bridge project was a union job even though the steel provider, Phoenix Bridge, had a long and somewhat successful history of refusing to work with unions. The larger union efforts of the IAB&SIW, however, were severely hurt by the dynamiting strike of the early 1900s that peaked with the explosion of the *Los Angeles Times* building in 1910 that ruined the union's public credibility. The structural steel workers who erected the Manhattan Bridge were no doubt helped by the emergence of a union and by the demands of insurance companies to introduce cages and other safety equipment. No extraordinary accidents spoiled this project, but nothing could alter the nature of the essentially dangerous work that the workers performed.

The first forty years of the bridge's service to New Yorkers were in most respects relatively quiet decades in the life of the new bridge. *The New York Times* reported on some litigation in the wake of construction that hinted of payoffs and corruption, but there was no major scandal. Not surprisingly, auto and truck traffic grew almost geometrically through the 1920s and even the 1930s, and this meant that horse-drawn vehicles would have to be excluded from the new span. Some significant repair work was carried out in the late 1930s, but this was not perceived as a foreshadowing of major structural defects. The Second World War raised security issues, and convinced some planners that tunnels would be safer than bridges in the future. By the late 1940s the Manhattan Bridge had become a recognized public monument that not only served both commuters and "day trippers" to Coney Island effectively, but provided a magnificent bonus in the visual panorama that included the East River, other bridges, and the Manhattan skyline.

Between 1950 and 1978 major structural flaws became dreadfully apparent for the first time. Funding to repair or reconstruct the span had to come primarily from local and state sources; not surprisingly, work was carried out on a sporadic basis with no master plan in sight. Most shocking, perhaps, was the view of some in the mid-1950s that the Manhattan Bridge had no useful future and thus should be abandoned or torn down. This outlook may not seem totally outrageous considering the amount of money invested in the span between 1950 and 2000. The prevailing bias in the 1950s, however, was that commuters needed the bridge, that the structure possessed at least some artistic merit, and that it had become an established part of New York's landscape or riverscape, and thus should not be torn down the way elevated lines had been razed in Manhattan earlier in the twentieth century.

In the fifties, a citizen by the name of John Ward issued a provocative allegation, that major bridges had such a profound impact on the neighborhoods that housed their anchorages and approaches that the bridges themselves inadvertently became a source of falling land values and ultimately urban blight. While Ward was at it, he might as well have accused bridges of causing urban poverty—and he did! What extraordinary grist for the mills of urban economists and sociologists.

During the Carter administration the federal government made a commitment to finance the reconstruction of the substantial bridges of America. It was, finally, time to get serious about saving national treasures as well as enhancing the safety of the public.

The period 1978 through 2000 featured such fascinating subjects as Mayor Koch's 1980 proposal to ban all single-occupant vehicles from Manhattan (a serious political miscalculation), and President Reagan's controversial gasoline tax (met with a filibuster) that was designed to pay for the upgrading of the nation's highways and bridges. Most critical to the future of the Manhattan Bridge, however, was the beginning of a relationship with Weidlinger Associates of New York City, funded with federal dollars, that guaranteed that state-of-the-art civil-engineering expertise would be applied to reconstructing the span in a logical and consistent fashion over several years. At the same time, all of this planning can be made to look bad if the bridge begins to accelerate its rate of deterioration, or if conditions are worse than anyone could have predicted. For example, when civil engineers or others examine rust and corrosion on a steel plate or beam there is frequently good reason to imagine that there is good metal underneath or in back of the bad. Where the Manhattan Bridge is concerned, however, there have been too many instances of scraping off the rust and corrosion or removing a decaying piece of steel only to find the newly exposed steel to be in wretched condition as well. These are developments and disappointments that no one could be expected to foresee.

Herbert Rothman and Weidlinger Associates set to work in the 1980s and 1990s to see whether their reconstruction efforts could outpace accelerating deterioration and the poorly applied deflection theory of seventy years earlier. To that end, cracking and metal fatigue were attacked creatively through studying a 108-foot test section and ultimately utilizing torque tubes to reduce the wear generated by severe torque. The real key to saving the bridge, in the eyes of some, was to keep the brutally heavy subway trains off the span. Ironically, running the trains on the bridge and transporting a few hundred thousand riders daily was one of the few ways of justifying the expenditure of hundreds of millions of dollars to upgrade the structure.

To complicate the story, in the early 1990s the question of being earthquake-proof was raised. While clearly not as vulnerable as the Golden Gate Bridge for geological reasons, the New York metropolitan area does lie within a region that experiences some quakes. What level quake should the span be protected against, and at what cost?

During the 1990s the intensity and rate of repair has been staggering, as evidenced by the red and yellow warning flags assigned by inspectors. These flags on the Manhattan Bridge far outnumber those on any other span. Are the civil engineers winning the race against accelerating decay? Does continuing investment seem to be warranted? Is there an irrational love for the bridge that defies sound analysis? Stay tuned to this ongoing saga!

appendix

SCIENCE AND TECHNOLOGY INTERFACE

Some Insights From Thomas Kuhn, Paul Sibly, and Henry Petroski

To the extent that modern engineering schools have their students study the Manhattan Bridge, it is no doubt held up as a magnificent example of how not to design, build, maintain, and reconstruct a suspension bridge for heavy-duty use in a major metropolitan area. How did designers and engineers apparently fail in this instance? A useful analytical overview of part of the problem has been provided by Paul Sibly in a doctoral dissertation completed under Professor A.C. Walker at the University of London in 1977. Sibly's analysis drew heavily on Thomas Kuhn's seminal work, *The Structure of Scientific Revolution*.[1]

Professor Kuhn argued that a major shift in scientific thought, a new scientific paradigm, was achieved only with great difficulty over many years—frequently after a tradition-bound generation had retired or died off. Older and newer views of science competed in a Darwinistic struggle for survival, but older scholars balked at giving the new ideas a fair hearing, primarily because the older scientists could only see the world that their graduate mentors had carefully instructed them to see. Put another way, they were wearing "blinders," and thus were not as objective in their outlook as they might have imagined.[2]

Paul Sibly, in a moment of insight, saw how Kuhnian thought shed light on the history of bridge building. He identified a preference in bridge type or design that constituted the one best way of doing something (a governing paradigm) that seemed to endure for roughly a generation or thirty years. Over the three decades the new design would be refined and refined again. More and more steel would be removed to save money and to provide the span with a sleeker look. The basic design was reconceived and reconceived, argued Sibly, until it actually became dangerous — even though no one may have understood this. In due time a major disaster occurred. At this point thoughtful people concluded that the old paradigm was obsolete and that a new one was needed. Accordingly, a new concept in bridge design superseded the old and a new paradigm emerged.[3]

To illustrate his point historically, Sibly started with the Dee Bridge disaster of 1847 in Chester, England. Here a span just over a hundred feet in length, constructed with cast-iron beams trussed with wrought-iron rods, failed under the weight of a train. The trusses that should have added strength to the bridge, according to Robert Stephenson's generation, alas, contributed to "lateral torsional instability" and thus collapse.[4] Moving beyond the failure of 1847, British designers strove to take more and more iron out of their truss bridges, and one result was Thomas Bouch's creation that spanned the extensive Firth of Tay in 1878. This elegant structure featuring "85 spans of wrought-iron lattice girders" certainly fell short of expectations in December of 1879 when several girders and a train with 75 passengers toppled into the firth. According to Petroski, the girders enveloping the train had been driven off their supports by high winds.[5]

The Tay Bridge episode sparked intensive interest in alternatives, particularly in cantilevered bridges. To that end, John Fowler designed a "massive" cantilevered bridge, the first in the world to have a steel superstructure, to span the Firth of Forth. Completed in 1890, each member was so awesome in its tubular appearance that the naïve observer could easily believe each member could support ten times the weight it was actually expected to bear. Surely a new paradigm had been established: thus the Forth Bridge became the inspiration for the now-infamous first Québec Bridge, built over the St. Lawrence River just outside of Québec City in the early 1900s. The existing faith in this new model was so powerful that the Québec authorities dismissed the distinguished designer Gustav Lindenthal and his design for a suspension bridge over the St. Lawrence. On a project that was short of funds, they paid him $4,500 to go away and substituted plans for their own cantilevered span. Before this structure could be completed, on August 29, 1907, the southern arm and 83 men crashed into the river, killing 75. In essence, the buckling of a strut, which consulting engineer Theodore Cooper and others chose to ignore, convinced a new generation that cantilevered bridges were not the hope of the future.[6]

At the time of the Québec tragedy the Manhattan Bridge was being erected, and the grand new concept of that era among civil engineers was deflection theory. The emphasis here was on relatively lightweight suspension bridges that would certainly flex and to some degree sway in the breeze. The application of "early deflection theory" to Manhattan Bridge helps to explain never-ending rebuilding and reconstruction. When deflection theory was carried to the nth degree in 1940 with erection of the Tacoma Narrows span, the stage was set for another disaster—a modern calamity actually captured on film. With the Tacoma Narrows story came the need for yet another paradigm. And so it goes.

In the preceding examples Paul Sibly has shown how Kuhnian theory from the field of science can be applied to explain developments in the history of engineering. Sibly's application cogently demonstrates how generations of engineers moved from the accepted wisdom of one era to the accepted wisdom of the next, with spectacular disasters generally serving as the trigger mechanism. An elegant interface between science and technology!

epilogue

On the morning of September 11, 2001, two hijacked commercial airliners crashed into World Trade Center towers on the west side of lower Manhattan, thereby creating a disaster heretofore anticipated only in a Tom Clancy novel. The resulting explosions and collapsing towers hurled tons of lacerating shrapnel and jagged debris as far as the Lower East Side, where some terrified citizens took refuge behind and under the East River bridges. When the air seemed to clear of flying projectiles, but not smoke, tens of thousands of workers from the financial district exited the neighborhood by walking across the Brooklyn, Manhattan, and Williamsburg spans to the relative safety of Brooklyn. In the face of a high-tech disaster in a high-tech age, three century-old suspension bridges provided some of the few options for escaping lower Manhattan.

In the eerie aftermath of a tragedy in which several thousand people had been incinerated or vaporized, such that the remains of many would never be recovered, these same antiquated spans served as vital arteries for a wide variety of emergency vehicles. Private automobiles were not permitted back on the East River bridges between 6 a.m. and 12 noon until Thursday, September 27, 2001—and then only with at least one other passenger. There is considerable irony in noting that when America's largest city confronted, arguably, its greatest crisis, three stately and decaying architectural icons spanning the East River proved to be vital lifelines.

a note on sources

The business records of the Phoenix Iron and Steel Company and the Phoenix Bridge Company that fabricated and designed the steel for the Manhattan Bridge are housed in the archives of the Hagley Library in Greenville, Delaware. The Phoenix Bridge records shed little light on labor relations, but do contain Order Books, Erection Books, Insurance Records, and Engineers' Notebooks. The collection also includes Annual Reports and Board of Directors' Minutes.

A number of local newspaper stories relating to the works at Phoenixville are available at the Chester County Historical Society, West Chester, Pennsylvania. *The New York Times* was invaluable throughout the twentieth century. The newspaper of record was particularly good at tracking the history of repairs and reconstruction on Manhattan Bridge.

Several engineers provided a wealth of information and cooperation—from Herbert Rothman at Weidlinger Associates (responsible for restoration of the bridge) to John Kulicki at Modjeski and Masters in Mechanicsburg (responsible for the safety of the subways on the span); from Eric DeLony, Director of Historic America Engineering Record, National Park Service, to Professor David P. Billington of Princeton University; from Jagtar Kinda, currently analyzing the Manhattan Bridge on site, to Emory Kemp, Director of the Institute for the History of Technology and Industrial Archaeology in Morgantown, West Virginia. DeLony and Kemp read the entire manuscript.

A number of periodicals shed light on bridge construction projects. *Engineering News* and *Engineering Record* (same journal) was invaluable. *Scientific American* carried careful accounts of the debates over technical issues, punctuated with some observations on the influence that politics had in resolving technical problems. *Technology and Culture*, *IA* (Journal of the Society for Industrial Archaeology), and *American Heritage of Science and Technology* treat a variety of subjects relating to bridge history—as do the *Proceedings of the International Bridge History Conference*, held every few years.

Eric DeLony's articles in *IA*, Victor Darnell's *A Directory of American Bridge Companies, 1840–1900*, David Plowden's *Bridges*, Donald C. Jackson's *Great American Bridges and Dams*, David McCullough's *The Great Bridge*, and David P. Billington's books offer excellent background and perspective for anyone approaching bridge history for the first time.

The literature on New York City is, understandably, staggering, continues to grow, and need not be chronicled here. Robert A Caro's *The Power Broker: Robert Moses and the Fall of New York* stands out as an overview of transportation networks.

about the author

Thomas R. Winpenny is Professor of History at Elizabethtown College, Elizabethtown, Pennsylvania. He was educated in the public schools of Philadelphia. He earned a B.A. and M.A. from the Pennsylvania State University, and a Ph.D. in history from the University of Delware, where he was a Hagley Fellow.

Professor Winpenny has been teaching courses in American history, industrial history, and the history of technology at Elizabethtown since 1968. He serves as Associate Editor for the Economic and Business Historical Society, and recently served on the editorial board of Pennsylvania History. He has published numerous articles on industrial history and the history of technology, and is the author of three previous books, including *Without Fitting, Filing or Chipping: An Illustrated History of the Phoenix Bridge Company* (Canal History and Technology Press, 1996), which was the winner of an Editor's Book Award from Gordon and Breach Publishers.

The author resides in Lancaster and is married to Marilyn Sims Winpenny.

Professor Winpenny may be reached by e-mail at winpentr@etown.edu.

notes

FOREWORD

1. Leon Solomon Moisseiff was born in Riga, Latvia, and studied at the Baltic Polytechnic Institute in Riga before emigrating to New York City with his family in 1891. There he studied civil engineering at Columbia University (1892–95) and worked with the New York Department of Bridges (1898–1910) where he was a design engineer for the Williamsburg, Manhattan, and Queensboro bridges. As an independent consulting engineer (1915–1940), he was involved with suspension bridge projects including the Delaware River Bridge (Philadelphia); the George Washington, Triborough, and Bronx- Whitestone bridges (NYC). He was design engineer for the Tacoma Narrows (Galloping Gertie) in Washington. Although blamed for the wind-driven collapse of the Tacoma Narrows Bridge in 1940, Moisseiff was still so esteemed by the engineering profession that the American Society of Civil Engineers established the Moisseiff Award that is awarded annually for an important paper published by the Society dealing with the broad field of structural design, including applied mechanics, as well as the theoretical analysis or improvement of such engineering structures as bridges and frames. Moisseiff himself published numerous influential papers on the design and construction of long-span bridges.
2. J.K. Finch. "A Hundred Years of American Civil Engineering, 1852–1952," *American Society of Civil Engineers, Centennial Transactions*, Volume CT, 1953, 61.
3. http://www.nycroads.com/crossings/manhattan/
4. "Manhattan Bridge," *Weidlinger Associates Newsletter,* New York City, spring 1996.
5. Earl E. Dubin, PE, Structural Engineer, Federal Highway Administration, New York Division, Bojidar S. Yanev, Eng.Sc.D., PE, Director, Bridge Inspection & Bridge Management, Department of Transportation, New York City, "Managing the East River Bridges in New York City" (Washington, DC: Federal Highway Administration, 1999).
6. Leon Moisseiff, "Growth of Suspension Bridge Knowledge," *Engineering News-Record 123,* 17 August 1939, 206–207.
7. Stephen Buonopane and David Billington, "Theory & History of Suspension Bridges Design from 1823 to 1840," *Journal of Structural Engineers,* Vol. 119, No. 3, March 1993 (New York: American Society of Civil Engineers), 963–967.
8. These remarkable advances in the theoretical design of suspension bridges are from an upbeat article written by Moisseiff entitled "Growth in Suspension Bridge Knowledge," and published in the 17 August 1939 issue of *Engineering News Record*, just a few months before the Tacoma Narrows failure.
9. James W. Doig and David P. Billington, "Ammann's First Bridge: A Study in Engineering, Politics, and Entrepreneurial Behavior," *Technology & Culture*, 1994, 554.
10. Othmar H. Ammann, "The Present Status of Design of Suspension Bridges with Respect to Dynamic Wind Action," *Journal of the Boston Society of Civil Engineers* 40, 1953, 236.

11. Buonopane and Billington, 967.
12. "Credit Where Credit Was Due," reprinted from Purdue Engineering *Extrapolations*, summer 2001.
13. Richard Scott. *In the Wake of Tacoma: Suspension Bridges in the Quest for Aerodynamic Stability* (Reston, Virginia: ASCE Press, 2001), 68.
14. Buonopane and Billington, 954.
15. F.B. Farquharson, F.C. Smith, and G.S. Vincent, *Aerodynamic Stability of Suspension Bridges with Special reference to the Tacoma Narrows Bridge: A Report of an Investigation Conducted by the Structural Research Laboratory, University of Washington, Bulletin 116, in IV Parts* (Seattle: University of Washington Press, 1949–1954); Kranakis, 155.
16. Farquharson report, 13.
17. Scott, 95.
18. *ENR*, May 1941, 77, 79.
19. Scott, 114 115.

CHAPTER ONE

1. Thomas Kuhn, *The Structure of Scientific Revolution* (Chicago: University of Chicago Press, 1962).
2. David McCullough, *The Great Bridge* (New York: Simon and Schuster, 1983).
3. Sidney Fine, *Without Blare of Trumpets* (Ann Arbor: University of Michigan Press, 1995).
4. Tom Kelly, MTA spokesman, quoted in *The New York Times*, 11 Sept 1997.

CHAPTER TWO

1. For a discussion of the nexus between bridges, transportation networks, social needs, and ethnic population shifts see Sharon Reier, *The Bridges of New York* (New York: Quadrant Press, Inc., 1977).
2. For a discussion of the planning designed to meet the transportation needs of Lower Manhattan (together with a map) see *Engineering News*, Vol. L, #27, 31 Dec 1903, 594.
3. The dynamite strike carried out by the IAB&SIW is carefully described and analyzed by Sidney Fine in *Without Blare of Trumpets: Walter Drew, the National Erectors' Association, and the Open Shop Movement, 1903–1957* (Ann Arbor: University of Michigan Press, 1995). While most explosions involved structural steel that had been stockpiled on a job site, ultimately the offices of *The Los Angeles Times* became a target—with deadly results.
4. For a thorough discussion of the career of Gustav Lindenthal see Chapter IV of Henry Petroski's *Engineers of Dreams, Great Bridge Builders and the Spanning of America* (New York: Alfred A. Knopf, 1995).
5. "The Budapest and Manhattan Suspension Bridges," *Scientific American*, 9 Sept 1905, 194. See also two additional articles in the same journal for 4 Nov 1905.
6. See *Engineering News*, *Scientific American*, and *The New York Times*, for the years 1902–1905.
7. *Engineering News*, Vol. L, #7, 13 Aug 1903, 142.

8. Clifford W. Zink and Dorothy White Hartman, *Spanning the Industrial Age: The John A. Roebling's Sons Company, Trenton, New Jersey, 1848–1974* (Trenton: Trenton Roebling Community Development Corporation, 1992), 89.
9. *Engineering News*, Vol. L, #27, 31 Dec 1903.
10. *The New York Times*, 8 May 1904, 6.
11. Ibid., 15 July 1903, 14.
12. Ibid.
13. *Scientific American*, 6 May 1905.
14. See "Not Quite Spanning the St. Lawrence: The Dreadful and Famous Québec Bridge Disaster of 1907" (Chapter VI) of Thomas R. Winpenny, *Without Fitting, Filing, or Chipping: An Illustrated History of the Phoenix Bridge Company* (Easton: Canal History and Technology Press, 1996).
15. Board of Directors, *Minutes*, Phoenix Bridge Company, 9 Oct 1907, Phoenix Bridge Collection, Hagley Archives, Greenville, Delaware. A reasonably thorough (though not complete) collection of corporate records for Phoenix Iron and Steel and their subsidiary the Phoenix Bridge Company are to be found in the Hagley Archives.
16. See David Plowden, *Bridges: The Spans of North America* (New York: Viking Press, 1974), 180–181.
17. Contract between Ryan-Parker Construction Company and the Phoenix Bridge Company dated 10 Oct 1906 and 17 Nov 1906, Phoenix Bridge Collection, Hagley Archives.
18. Even if these estimates could not be relied upon as something approaching absolute truth, they were startling enough to settle the issue.
19. Edwin T. Layton in a review of Petrowski's *Engineers of Dreams* for the *American Scientist,* July-August 1996. For additional insights see Jameson W. Doig and David P. Billington, "Ammann's First Bridge: A Study in Engineering, Politics, and Entrepreneurial Behavior," *Technology and Culture*, Vol. 35, No. 3, July 1994, 537–570, and Stephen G. Buonopane and David P. Billington, "Theory and History of Suspension Bridge Design from 1823 to 1940," *Journal of Structural Engineering*, Vol. 119, No. 3, March 1993, 954–977.
20. "Manhattan Bridge," insert in *Weidlinger Associates Newsletter*, spring 1996. Weidlinger Associates is a civil engineering firm in New York City.
21. *History of Orders*, 16 Aug 1906 through 5 Jan 1910 in Phoenix Bridge Company Collection, Hagley Archives.
22. *The New York Times*, 28 Feb 1909.
23. *Index of Orders*, 1906–1910, Phoenix Bridge Company Collection, Hagley Archives.
24. Estimates garnered from *Scientific American*, 5 Dec 1908, 398–399. Arguably, any such estimates are grossly oversimplified.
25. *The New York Times*, 30 Jan 1910, 3.

CHAPTER THREE

1. *The Bridgemen's Magazine*, Oct 1907, 687, and Dec 1909, 871–872, contains data on membership. This labor periodical, established in 1896, is available from the Center for Research Libraries, 6050 Kenwood Avenue, Chicago, Illinois 60637.
2. *The Bridgemen's Magazine*, 1896–1909.
3. See Sidney Fine, "The National Erectors' Association and the Dynamiters," *Labor History* Vol. 32, No. 1, winter 1991, 5.

4. Ernest Poole, "Cowboys of the Skies," *Everybody's Magazine*, 1908.
5. Ernest Poole, "Cowboys of the Skies," *The Bridgemen's Magazine*, February, 1909.
6. Ibid., 72.
7. Ibid., 73.
8. Ibid., 77.
9. *The Bridgemen's Magazine*, Nov 1908, 657.
10. Ibid., Oct 1907, 632.
11. Tom Buckley, "A Reporter at Large: The Eighth Bridge," *The New Yorker,* 14 Jan 1991, 50.
12. *The Bridgemen's Magazine*, Oct 1908, 721.
13. See *The Bridgemen's Magazine*, Sept 1909, 499, and Buckley, "Reporter at Large."
14. *The New York Times*, 10 Jan 1909, 3:2.
15. Ibid., 13 April 1909, 4:5.
16. Ibid., 17 July 1909, 5:5.
17. Ibid., 20 Aug 1909, 14:3.
18. *The Bridgemen's Magazine*, Jan through Dec 1909.
19. Luke Grant, *The National Erectors' Association and the International Association of Bridge and Structural Iron Workers* (Washington, D.C.: United States Commission on Industrial Relations, 1915), 8.
20. *Bureau of Labor Statistics, Bulletin 234* (Washington, D.C.: U.S. Government Printing Office, 1918), 241.
21. *The Bridgemen's Magazine* Oct 1907, 728.
22. See Thomas R. Winpenny, *Without Fitting, Filing, or Chipping: An Illustrated History of the Phoenix Bridge Company* (Easton: Canal History and Technology Press, 1996), Chapter 6.
23. *The Bridgemen's Magazine*, Oct 1907, 728.
24. See Winpenny, *Phoenix Bridge*, Chapter 5.
25. See *The Bridgemen's Magazine*, Oct 1907.
26. Sidney Fine, *Without Blare of Trumpets: Walter Drew, the National Erectors' Association, and the Open Shop Movement, 1903–1957* (Ann Arbor: University of Michigan Press, 1995), 47.
27. Ibid., 39.
28. Ibid., 42.
29. See Patrick Renshaw, *The Wobblies: The Story of Syndicalism in the United States* (Doubleday: Garden City, New York, 1968).
30. See Winpenny, *Phoenix Bridge*, Chapter 7.
31. *The New York Times*, 27 April 1908, 2:6.
32. Fine, "The National Erectors' Association," *Labor History*, 7.

33. *The Bridgemen's Magazine*, Oct 1908, 726.
34. John Sterling Deans to O.F. Nichols, 18 Aug 1906, Chief Engineer's Correspondence, Phoenix Bridge Company Collection, Hagley Archives.
35. *The Bridgemen's Magazine*, April 1908, 208.
36. See John Sterling Deans, *Correspondence.*
37. John Sterling Deans to Ryan-Parker Construction, 28 Jan 1907.
38. Ibid., 15 March 1907.

CHAPTER FOUR

1. See George J. Lankevich, *American Metropolis: A History of New York City* (New York: New York University Press, 1998), chapter 7.
2. Brian Cudahy, *Under the Sidewalks of New York: The Story of the Greatest Subway System in the World* (New York: Fordham University Press, 1995), 66.
3. Ibid.
4. *The New York Times*, 26 April 1912, 1:5.
5. Ibid., 10 Feb 1912, 20:1.
6. Ibid., 3 July 1919, 5:1.
7. John F. Kasson, *Amusing the Millions: Coney Island at the Turn of the Century* (New York: Hill and Wang, 1978).
8. *The New York Times*, 7 July 1922, 1:5.
9. Ibid.
10. Lankevich, *American Metropolis*, 259.
11. *The New York Times*, 24 May 1926, 3:5.
12. See *The New York Times* between January and June of 1929.
13. *The New York Times*, 10 Nov 1929, 19:2.
14. U.S. Bureau of the Census, *Bicentennial Statistics*, U.S. Department of Commerce, 1976.
15. *The New York Times*, 23 Jan 1932, 3:6.
16. Ibid., 6 June 1931, 33:4.
17. Ibid., 23 April 1931, 27:6.
18. Ibid., 1 Jan 1935, 22:7.
19. Ibid., 24 Aug 1935, 17:6.
20. Ibid., 7 April 1937, 27:1.
21. Ibid., 27 March 1938, II, 1:7.
22. Ibid., 18 July 1939, 1:3.

23. Ibid., 17 Dec 1941, 32:2.
24. Ibid., 24 July 1942, 13:4.
25. See *The New York Times* between 1946 and 1950.

CHAPTER FIVE

1. Roger Kahn, *The Boys of Summer* (New York: Harper-Collins, 1987).
2. *The New York Times*, 27 April 1952, 76:3.
3. Ibid.
4. Ibid.
5. Ibid.
6. Ibid., 25 May 1952, 48:1.
7. Ibid., 14 May 1953, 24:3.
8. Ibid., 13 Aug 1953, 27:5.
9. Ibid., 25 Aug 1953, 20:7.
10. Ibid.
11. *Scientific American*, 15 June 1878.
12. Benson Bobrick, *Labyrinths of Iron: A History of the World's Subways* (New York: Newsweek Books, 1981), 205.
13. For a classic and controversial academic discussion of this issue, see Edward C. Banfield, *The Unheavenly City Revisited* (Boston: Little, Brown, 1974).
14. *The New York Times*, 1 Sept 1953, 22:5.
15. Ibid., 29 July 1954, 25:2.
16. Ibid., 27 Jan 1955, 15:5.
17. Ibid., 11 Feb 1955, 12:3.
18. Ibid., 16 March 1955, 38:8.
19. Ibid., 25 Oct 1959, 56:3.
20. Ibid., 2 Jan 1959, 23:4.
21. Ibid.
22. *The New York Times*, 25 June 1958, 1:8.
23. Ibid., 25 Oct 1959, 56:3.
24. Robert Caro, *The Power Broker* (New York: Knopf, 1974), 9.
25. Ibid., 5.
26. *The New York Times*, 1 June 1961, 37:2.

27. Ibid., 2 June 1961, 22:4; 7 June 1961, 40:2.
28. Ibid., 7 July 1961, 27:3.
29. Ibid., 13 April 1964, 31:1.
30. Ibid., 24 Nov 1970, 81:2.
31. Ibid., Aug 1975 through Aug 1976.
32. Ibid., 11 Jan 1972, 1:1.
33. Ibid., 19 Aug 1977, 1:1.

CHAPTER SIX

1. *The New York Times*, 27 March 1978, B:1.
2. Ibid., 20 March 1979, II, 1:5.
3. Herbert Rothman, "The Manhattan Bridge Story," *Weidlinger Associates Incorporated Newsletter*, spring 1996–insert: http://www.wai.com/Newsletter/Spring96-insert/wanews1.html
4. *The New York Times*, 3 July 1980, 1:2.
5. Ibid., 24 July 1980, II, 3:5.
6. Clark Wieman, "Road Work Ahead: How to Solve the Infrastructure Crisis," *Technology Review*, Jan 1993, 43.
7. *The New York Times*, 30 June 1981, II, 1:1.
8. Ibid., 1 Nov 1981, 41:1.
9. Ibid., 26 May 1982, II, 3:6.
10. Ibid., 24 Nov 1982, 1:6 and 24 Dec 1982, 1: 6.
11. Ibid., 11 June 1983, 12:4.
12. Ibid., 4 Aug 1983, 1:1.
13. Rothman, *Weidlinger Newsletter*, 1.
14. Ibid., 2.
15. Ibid.
16. Ibid., 5.
17. Ibid., 6.
18. Ibid., 4.
19. Ibid., 7.
20. David Berreby, "The Great Bridge Controversy," *Discover*, Feb 1992, 33.
21. Rothman, *Weidlinger Newsletter*, 8.

22. See Calvin Sims, "Corrosion's Rise Imperils Bridges Over East River," *The New York Times*, 17 Dec 1990, 1:1, and a host of additional stories throughout the 1990s.
23. *The New York Times*, 17 Dec 1990, 1:1.
24. Ibid.
25. Ibid.
26. *The New York Times*, 8 Jan 1991, 1:1.
27. Calvin Sims, "Badly Deteriorated Manhattan Bridge," *The New York Times*, 15 Feb 1991, 2:4.
28. Ibid.
29. Calvin Sims story, *The New York Times,* 27 Feb 1991, 1:8.
30. "Ammunition for A Hearing on Mismanagement," *The New York Times*, 28 Feb 1991, 1:8.
31. *The New York Times*, 31 July 1992, B, 1:1.
32. Ibid., 5 June 1994, II, 24:1.
33. Ibid., 2 May 1995, B, 3:1.
34. Ibid., 10 April 1996, B, 1;2.
35. Ibid., 11 Sept 1997, B, 3:2.
36. Bonnie Yochelson, "A Fantastic Passion for New York," http://www.mcny.org/abbott/passion.htm

CHAPTER SEVEN

1. See Horst Hamman, *New York Vertical (Cult Cities of the World)* (te Neues Publishing Company, 1998), and Stanley Greenberg, *Invisible New York* (Baltimore: Johns Hopkins University Press, 1998).
2. Carrere and Hastings was an architectural firm located in New York City.
3. It is unlikely in the first decade of the twentieth century that the circulation of *The Bridgemen's Magazine* was much greater than 10,000. It also seems reasonable to conclude that "readership" was not a good deal higher than circulation.

APPENDIX

1. Thomas Kuhn, *The Structure of Scientific Revolution* (Chicago: University of Chicago Press, 1962).
2. Ibid.
3. Paul Sibly, "The Prediction of Structural Failures," Ph.D. thesis, University of London, 1977.
4. Ibid.
5. Henry Petroski, "Predicting Disaster," *American Scientist*, Vol. 81 (March/April 1993), 110-113, 110.
6. For a discussion of the accident see Thomas R. Winpenny, *Without Fitting, Filing, or Chipping: An Illustrated History of the Phoenix Bridge Company* (Easton, PA: Canal History and Technology Press, 1996), Chapter 6.

photo credits

The illustrations in this volume have been used with the permission of the following:

Hagley Museum and Library, Wilmington, Delaware. The Hagley archives contain the business records of the Phoenix Iron and Steel Company and the Phoenix Bridge Company.

Historic American Engineering Record, National Park Service.

The Museum of the City of New York, 1220 Fifth Avenue, New York City.

The National Canal Museum, Easton, Pennsylvania.

The New-York Historical Society, Two West 77th Street, New York City.

The Smithsonian Institution.

Thomas Flagg.

Front cover: Phoenix Bridge collection, © Hagley Museum and Library.

Back cover: Construction photograph, Phoenix Bridge collection, © Hagley Museum and Library; view from Manhattan shoreline, Historic American Engineering Record: Jet Lowe, photographer, 1979. HAER, NY,31-NEYO,164-7.

xi Historic American Engineering Record: Jack Boucher, photographer, 1978. HAER, NY,31-NEYO,165-;DLC/PP-97:NY-14

xii Historic American Engineering Record: Jet Lowe, photographer, 1986. HAER, NY,31- NEYO, 161-14.

xiv Historic American Engineering Record: Jet Lowe, photographer, 1995. HAER, CAL,38-SANFRA,141-:DLC/PP-02:CA-34.

xv Historic American Engineering Record: Jet Lowe, photographer, 1984. HAER, CAL,38-SANFRA, 140-2.

xvi Historic American Engineering Record: historic view, 1940. HAER, WASH,27-TACO,11-31.

xx Artist's view of Manhattan Bridge from the Brooklyn side, showing how the bridge would look when completed. Trains are crossing on each side. An armored cruiser, possibly the USS New York, is underneath the bridge. *The Manhattan Bridge under construction.* Half-tone, undated. Museum of the City of New York Print Archives (©MCNY).

11 *Scientific American*, NCM collections.

16 *Scientific American*, NCM collections.

17 Phoenix Bridge collection, © Hagley Museum and Library.

19 Phoenix Bridge collection, © Hagley Museum and Library.

20 Collection of The New-York Historical Society, digital i.d. number 75947, © The New-York Historical Society.

24 Collection of The New-York Historical Society, digital i.d. number 75948, © The New-York Historical Society.

25 Collection of The New-York Historical Society, digital i.d. numbers 75949 (left) and 75950 (right), © The New-York Historical Society.

28 *Manhattan Bridge Pier.* View from Water Street and Mechanics Alley. Museum of the City of New York Print Archives (©MCNY).

32 Collection of The New-York Historical Society, digital i.d. number 75951, © The New-York Historical Society.

33 Collection of The New-York Historical Society, digital i.d. number 75952, © The New-York Historical Society.

34 *Progress on the Manhattan Bridge.* March 27, 1909. Museum of the City of New York (©MCNY).

38 Collection of The New-York Historical Society, digital i.d. number 75953, © The New-York Historical Society.

41 Photo: Bernard Brown. *Manhattan Bridge*. General view of the plaza at the Brooklyn end of Manhattan Bridge. Undated. Museum of the City of New York. Gift of Photographer. (©MCNY)

43 *Jimmy Walker in a Parade.* 1925-1935. Museum of the City of New York. Estate of Louis J. Gribetz. 98.170.6. (©MCNY)

44 *Manhattan Bridge and East River from roof of 40 Monroe Street, Manhattan, corner Market Street known as Knickerbocker Village.* Jan. 1940. Museum of the City of New York. Gift of Mr. Bernard Kardon at the request of Museum Photographers Association. (©MCNY)

46 Berenice Abbott. *Manhattan Bridge.* November 11, 1936. Museum of the City of New York. Museum Purchase with funds from The Mrs. Elon Hooker Acquisition Fund. 40.140.164 (#185). (©MCNY)

48 Thomas Flagg Collection.

51 National Museum of American History, #93-5808, Smithsonian Institution.

53 Berenice Abbott *Oyster Houses, South Street and Pike Slip*. April 1, 1937. Museum of the City of New York. Federal Arts Project, "Changing New York" 40.140.189 (#232). (©MCNY)

59 *Robert Moses in Front of Map,* undated. Museum of the City of New York Print Archives. (©MCNY)

60 Collection of The New-York Historical Society, negative number 58802, © The New-York Historical Society.

61 Collection of The New-York Historical Society, digital i.d. number 75954, © The New-York Historical Society.

64 Historic American Engineering Record: Jack Boucher, photographer, 1978. HAER, NY,31- NEYO, 164-1.

66 Historic American Engineering Record: Jack Boucher, photographer, 1978. HAER, NY,31- NEYO, 164-4.

77 Historic American Engineering Record: Jet Lowe, photographer, 1991. HAER, NY,31- NEYO,164-8.

79 Thomas Flagg Collection.

80 Thomas Flagg Collection.

82 Historic American Engineering Record: Jet Lowe, photographer, 1982. HAER, NY,31- NEYO,90-2.

index

N

O

P

Q

R

S

T

U

V

W

Y

Z